Chinese Tourism Research Annual 2008

Tourism Tribune（2006–2007）English Edition

中国旅游研究年刊·2008

——《旅游学刊》（2006~2007）英文版

Editorial Department of *Tourism Tribune*

《旅游学刊》编辑部编辑

社会科学文献出版社
SOCIAL SCIENCES ACADEMIC PRESS (CHINA)

Tourism Tribune

Sponsor Institute of Tourism, Beijing Union University

Honorary Editor Liu Deqian

Editor-in-Chief Zhao Peng

Associate Editor Song Zhiwei   Wu Qiaohong

Editor Council (in alphabetical order)

Feng Dongming	Lian Yuejuan	Luo Xuhua
Ning Zequn	Song Zhiwei	Song Ziqian
Wang Bing	Wang Meiping	Wang Yujie
Wu Qiaohong	Zhao Li	Zhao Peng
Zhu Xiyan		

Director of Editorial Department Song Zhiwei

Vice Director of Editorial Department Wu Qiaohong

旅游学刊

主　　　　办　　北京联合大学旅游学院

名　誉　主　编　　刘德谦

主　　　　编　　赵　鹏

副　主　编　　宋志伟　吴巧红

编　　　　委　　冯冬明　廉月娟　罗旭华　宁泽群

宋志伟　宋子千　王　兵　王美萍

王玉洁　吴巧红　赵　丽　赵　鹏

朱锡炎

编　辑　部　主　任　　宋志伟

编　辑　部　副　主　任　　吴巧红

Preface

We are very pleased to offer you, our international readers, this *Chinese Tourism Research Annual 2008*. This is the third English version of *Tourism Tribune* about China tourism research after we published *Annual 2005* and *Annual 2006*. The 20 selective articles in this book are collecting from the 400 articles published on *Tourism Tribune* during 2006 and 2007.

It has been 20 years since its first publication, and *Tourism Tribune* has long been regarded as the earliest professional periodical on tourism research enjoying overseas and domestic readers. It is the most authorized periodical in terms of academic value since it is the only one that has been officially awarded for four times as Centered Periodical in China's tourism research.

We are happy to know that the success of tourism research in China is no less than that in foreign countries. For this reason, we published this English version of *Tourism Tribune* to help foreign scholars know more about the development of tourism research in China.

The scholars who selected articles for this book are aware of the importance of their selections, and understanding the needs of the foreign readers, so they paying large attentions to the academic value of the articles as well as the subjects coverage of the research.

We wish this English version would play a positive role in improvement of international tourism research and academic exchange. This is also the common wishes of all the people who have provided great assistance for the growth of *Tourism Tribune*, also, the wishes of our friends and Institute of Tourism, Beijing Union University.

Editorial Department of *Tourism Tribune*
August 2008

前　言

　　很高兴,《中国旅游研究年刊·2008》和大家见面了。这是我们继 2005、2006 年编辑出版《中国旅游研究年刊》的第三本中国旅游研究的优秀论文集。本书这里收入的 20 篇中国旅游研究的学术论文,是从《旅游学刊》2006、2007 年发表的中国作者的近 400 篇论文中精选出来的。

　　《旅游学刊》创刊至今已经 20 多年了,一直是读者公认的中国最早的海内外公开发行的旅游科学的专业性学术期刊。尽管近 10 余年中国的旅游类刊物已雨后春笋般地大量出现,但在中国图书信息系统的近 10 余年连续 4 次的全国"中文核心期刊"的郑重评定中,《旅游学刊》却是中国唯一的连续 4 次获得"中文核心期刊"荣誉的旅游类期刊。所以在中国,在中国的学术界和中国的旅游业者中,以及在部分海外地区的读者中,《旅游学刊》已经是人们一致公认的最具权威性的刊物。

　　20 多年米,中国的旅游科学研究已经取得了十分可观的成绩,为了有利于全球的这一领域的学术交流,在海外朋友们的建议下,我们特意对《旅游学刊》进行精选,并安排隔年连续出版英译本《中国旅游研究年刊》。

　　参与本书遴选的专家在推荐和挑选论文时,既注意了论文在学科成就方面的代表性,注意了全部论文所形成的对中国旅游研究的覆盖面,而且也注意入选论著的作者的一定代表性和在中国各地的覆盖面,因为他们都十分明白此书编辑出版的意义和价值。

　　但愿本书的编辑出版能对旅游科学的国际交流发挥出它应有的作用。这也就是所有为此付出辛勤劳动的专家学者、所有积极关心支持这一工作的单位和个人——北京联合大学旅游学院和所有朋友们的共同心愿。

《旅游学刊》编辑部

2008 年 8 月

CONTENTS

目　录

Tourist Experience from the Perspective of Phenomenology: Tourist World and Life—world

Xie Yanjun[1], Li Miao[2]

(1. School of Tourism & Hotel Management, Dongbei University of Finance & Economics, Dalian 116025, China; 2. School of Technology and Vocation, Dongbei University of Finance & Economics, Dalian 116025, China)

Abstract: This paper discusses some basic conceptions in the sense of theoretical studies. According to the points of view of the author, tourism phenomenon is something complicated with a hard core of tourist experience. Such an experiential process is a successive self–organizational system consisting of some unique and meaningful situations or settings which set up an alternative behavioral environment for tourists. The author defines this environment as tourist world and describes its structure further from the perspective of phenomenology.

Key words: tourist experience; tourist world; tourist setting; phenomenological methodology

As for tourism, people have formed some basic conceptual cognition based on common sense. They realize that though tourists travel for different purposes, they have something in common. That is, tourists travel to a place away from

[About authors] Mr. Xie Yanjun, Ph. D., is a professor and the dean of the School of Tourism & Hotel Managemen, Dongbei University of Finance & Economics. His research interests have been centered on basic theoretical studies of tourism. Now he is conducting research on tourist experience, socio-cultural impacts of tourism and tourism research methodology; Ms Li Miao, is a lecturer in the School of Technology and Vocation, Dongbei University of Finance and Economics. Her research interest is about the study of tourism management.

home but will finally return; they spend money and most importantly, a whole period of free time which is different from working hours. Due to these apparent characteristics, tourists' behaviour and its meanings are possible to be greatly different from those in daily environment. It is this possibility that gives us the impetus to make further research on tourist experience.

In order to found a solid logical framework for our exploration, the first question we are confronted with is to tell the differences between tourist phenomena and daily life phenomena. This will help reveal the features of tourism, thus make it possible to disclose the laws underlying tourism and tourist experience.

1. The formulation of a theoretical model: tourist world and daily life world

We consider Daily Life World to be the daily world where potential tourists live. It comprises the sum of a potential tourist's involvement in everyday affairs, with the only exception of affairs in tourist world (at most, these affairs may be partially included in or overlap with everyday affairs). Here we have to make a presupposition that tourism is an experiential process different from everyday life. Apparently, in a man's life-world, there may be (or are) affairs with more evident features than tourist affairs, however, because we are not interested in those affairs, we incorporate them into the mainstream of everyday activities and neglect their unique but irrelevant meaning. This made it possible for us to define a routine life world. This world is made up of everyday work, study, life and accidental affairs. And they may arouse certain emotions, such as boredom, detestation, shame, frustration and sorrow, and act as motives of tourism to a great extent.

Therefore, we can reason that tourist world is a completely new world different from the daily life world. Superficially, tourist world contrasts tensely with daily life world in two dimensions. Firstly, with respect to space, tourism world is always a temporary separation from daily life world. It is a departure and return from life-world. And during this period, changes will take place on tourists. Secondly, with respect to time, comparing with the whole (or full) time of life-world, time spent in places away from home will perish or leak out permanently. It will totally be a meaningless leakage if the whole meaning of a person's existence is only given by the life-world. Fortunately, in reality, the case is not true. People do not consider tourism as a pure time-killing practice. They find meanings in the process of tourism. Thus, time leaking out of life-world is subjectively and epistemologically meaningful.

It is the tourist experience that helps the realization of discovery and justification of a subject (namely a tourist). Tourist experience is a psychological process as well as a physical one; it is a time phenomenon as well as a space one; and it is an individual action as well as a social one. Tourist experience occurs in tourism world, thus it is within the limits of the tourism world. As for itself, with the shift of space and time, tourist experience brings psychological changes to the subject. The degree and direction of some changes may be made as expected, whereas others may be unexpected acquisitions. Whatever they are, tourists always manage their experiential process as a whole. They will make efforts to meet their expectations, or modify their experiential process to meet the basic part of initial expectations, or actively create and link some typical tourist settings to achieve their purposes of tourism. Hence, in the process of experience, tourists usually give the fullest play to their subjective initiatives, and they definitely do not deconstruct tourist experience into fragments, for that is not the purpose for tourism, neither its meaning (see the author's another article in journal of *Tourism Tribune* 20(4):8-9).

Taken in this sense, the process of tourist experience is such a successive self-organizational system consisting of some unique and meaningful settings. And tourists' behaviour orientation is greatly influenced by these settings. Hence, in order to forecast tourists' behaviour, it is necessary to understand those specific settings at first.

Following this way of thinking, we integrate these basic domains of life-world, tourist world, tourist experience and tourist setting into an inter-relative discourse system to explain the inner process and structure of tourist world. Their inter-relationship is illustrated in Figure 1. Next to this, we should further explore the structures and features of life-world and tourist world.

2. Natural attitude and intersubjectivity in life–world

Judging from orientation and rules of human's actions, our daily life is arranged within the institutional limits of social network of relationships. Even if the separate action of an individual may deviate from these restrictions, he is bound to be hedged in with culture. We live with purposes every day, and construct a life-world with unknown purposes. In this life-world, our human being is influenced by the external cultural and natural environment. Meanwhile, we experience difficulties and anxieties in it as a subjective self. Being enveloped in emotional atmosphere, human are experiencing the baptism of happiness and the

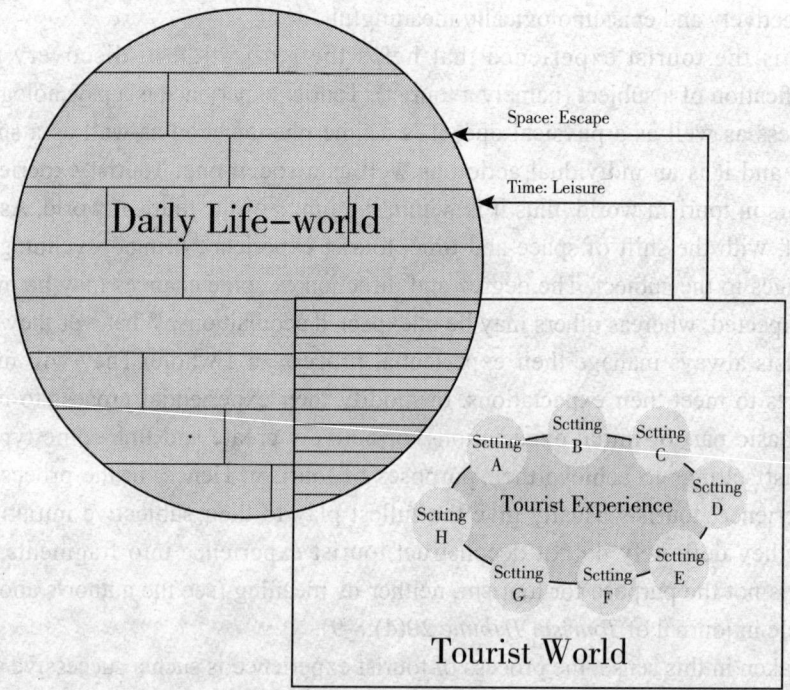

Figure 1   Relationship between Daily Life-world and Tourist World

torture of anxiety.

Introducing the concept of life-world or directly regarding life-world as the object of research is a viewpoint derives from phenomenology. Husserl asserted that life-world (lebenswelt) can give an obvious clue to the study of intention-in-action. In order to explore life-world and its structure, it is indispensable to perform an initial reduction or a suspension of scientific truth[①]. It is certain that reduction will take us from the structure of life-world back to its hidden intentional functions, then, these functions make it possible for us to investigate all features of life-world and the constitution of other objectivities based on them.

While Husserl defining the concept of life-world, he considered it to be

---

① Both reduction and epoché are important definitions in Husserl's phenomenology. Epoché is the negative aspect of reduction. Husserl speaks for the first time of "reduction" in his *Logical Investigations*. Negatively, it is a suspension of all judgments concerned with something transcendent (not immanently given). See De Boer, Th. *The Development of Husserl's Thought*. Trans. Th. Plantinga. The Hague: Nijhoff, 1978, p.308.)

a "world of lived experience"and different from daily life. The life-world is purposeless, while daily life can not work without purpose. Any of our practical worlds is different from life-world. On one hand, life-world works as the presupposition and basement of practical worlds; on the other hand, practical worlds constitute the life-world. Therefore, Husserl drew the conclusion that life-world is transcendental. In figure 1, we use patterns of circle and brick to represent the purposelessness and structure of life-world.

It is apparent that life-world is not an objective scientific world or the world in the sense of universe. It is a world that is experienced by a living subject from his specific viewpoint (no matter how seriously it is distorted). Therefore, obviously, life-world is a subjective and relative world. As for researchers, it is important to view life-world as an oriented world with a center which is labeled with the personal pronoun "self". Circling around the pole of "self", the world is made up of things such as "close" and "remote", "homeland" and "foreign country". Its spatial reference frame is experiencing to be stagnant, opposite to Copernicus' scientific paradigm[1].

Husserl's conception of life-world has inspired many researchers to observe and explore daily phenomena based on it. In the adoption of Husserl's "life-world", Alfred Schutz, one of the representatives of sociology and phenomenology, insisted that life-world as a cultural world is the base of human's daily activities, but he rejected the idea that life-world is transcendental. Hence, Schutz considers that "the life-world comprises the sum of man's involvement in everyday affairs".[2] Schutz's analysis of life-world closely relates with definitions of natural attitude and intersubjectivity. In his view, people living in life-world are characterized by natural attitude. Within the natural attitude, the reality of life-world is self-evident. Therefore, people in the natural attitude simply take life-world for granted. For ordinary men, life-world is taken for granted; at least most of the time, it is unquestionable. It has never been a "question" to draw our attention. On the contrary, this self-evident life-world presupposes any social action of ordinary men and their consideration on mundane "questions". In this sense, Schutz proposes that the natural attitude changes life-world from "our world" into "our world". After constructed by human action, life-world gradually externalizes into a world relative to us and in turn lays the foundation for ordinary men's action.

The reason why ordinary men's social action is based on life-world is that social action indicates communication with others and life-world is the presupposition of any social communication between others and me. In daily life, although everyone is posited in a unique space and time has his own perspective, life-world convinces us that different perspectives can be exchanged. I can think of a question from the standpoint of other people, for others are similar to me in life-world. We are all born in the same world, brought up from a child into a formal member of society following instructions of parents and other adults, learn the same language and communicate with fellow-men. Hence, our everyday knowledge of this world is "intersubjective", that is, the world is the same for all social members and in which reciprocal understanding is possible. Since actions of social members are based on this natural attitude and intersubjectivity, actors in life-world can reciprocally understand meanings of actions. Hence, they can easily deal with daily affairs. Therefore, in Schutz's view, life-world is different from the outer world in the eye of the general (especially natural scientists), because life-world is full of given meanings. The significance of the meanings is self-evident. It is difficult to make an appropriate reaction unless understanding them in the right way. Generally speaking, people can not give a serious explanation on the nature or connotative meaning of the life-world until they experience all kinds of elements and forms of daily life. Schutz future proposes that since an actor's subjective meaning plays a crucial role in the understanding of the action not for the purpose of intercourse, it will be indispensable for understanding actions with the purpose of intercourse (intercourse is the essential approach of human socialization)[3].

Therefore, here we can draw a clear conclusion: if we construct an everyday life-world, then, in this world, the existence of meaning is based on the agent's subjective intention and his action is restricted by his own subjective feature and environmental factors. In order to fully understand the meaning of an individual action, we should understand the agent's motives or subjective intention at first. Reciprocal understanding is possible because of the intersubjectivity and natural attitude in life-world. Further, it is only by such understanding of subjective intention that we can gain access to the meaning of action, and this is the right essence of phenomenology.

Nevertheless, how to relate everyday life-world with tourist world? Or what's the difference between them?

3. An independent tourist world escaped from life–world

According to what we discussed above, we know that for Husserl and Schutz, the term "life-world" is to be understood the scene of all our human activities, or "all activities directly participated in a person's living world" [4]. Life-world contains space experience, place experience, phenomenal experience and aesthetic (natural and cultural) experience. The fundamental connotative meanings of life-world are manifested by phenomenological method of understanding in its totality. Therefore, we need make denotative modification of Husserl's life-world. That is, the term "life-world" in our context is to be regarded as the everyday life-world relative to tourist world. In other words, if we keep the denotative and connotative meanings of Husserl's life-world, it is necessary for us to divide the world into two parts: everyday life-world and tourist world. Thus, if we view the life-world from the perspective of tourism, life-world can be regarded as a spatio-temporal continuum composing of everyday life-world and tourist world[1]. We'd better substitute the term of "everyday life-world" for "life-world" for convenience. Thus, in the course of our exploration, everyday life-world (life-world for short) is merely a world relative to tourist world.

Re-observing the everyday-life world from the perspective of tourism, what will it be like? Krippendorf (1986) once described everyday-life as "dirt, noise, work, rush, school, effort, spoilt environment, all these terms are included in 'everyday'. Even its description employs depressing terms and colours like grey, dull, tiring, sad, boring and so on"[5]. The reason why we can tolerate the long-term everyday routine is that we still have a dream to escape and travel far away. Tourism adds the spots of colour to our drab everyday canvas. They mean rejuvenation and renewal. We must travel away in order to prove our existence. We work to gain holidays, and we need these holidays to be able to work. When we regard everyday-life world in this way, the tourist world will consequently appear.

We can imagine that the process of tourism is meaningful with interlacement of space and time. As for anyone, if it is clear that his purpose is to go through tourist process in such space and time, hence it will be a process with a comprehensive structure. Tourist action seems to be open and structure-tensive. In fact, it is constructing a relatively closed world: when unfolded, it is the extension of life-world; when closed, it is a different world from life-world. We now can term this relatively closed and unique phenomenal space as tourist world. It has a complete system and a unique structure. This is a particular world in which tourist phenomena are different from those in life-world. As a tourist, he is living, thinking, feeling, moving and acting in this particular

---

[1]   In fact, during one's lifetime, tourism is not the only departure or escape from everyday life-worldlife-world. Since tourist experience is our main research subject, we can overlook all the other activities with different features from everyday life-worldlife-world. It means that we presuppose that the life-worldlife-world is constituted by two parts: everyday life-worldlife-world and tourist world.

world. Therefore, in tourist world, the experiences accumulated in life-world and questions based on these experiences will be nonexistent, or be changed in nature and form, or be changed in degree and direction. The reason is that a new standard of value judgment towards these questions will be established under the influence of a tourist's fundamental motive and presentation system.

In Urry's (2000) analysis of the elements of tourist gaze, the characteristics of tourism summarized by him are as follows[6]:

(1) Tourism is a leisure activity which presupposes its opposite, namely regulated and organized work. It is one manifestation of how work and leisure are organized as separate and regulated spheres of social practice in 'modern' societies.

(2) Tourist relationships arise from a movement of people to, and their stay in, various destinations. This necessarily involves some movement through space, that is the journeys, and periods of stay in a new place or places.

(3) The journey and stay are to, and in, sites outside the normal places of residence and work. Periods of residence elsewhere are of a short-term and temporary nature. There is a clear intention to return 'home' within a relatively short period of time.

(4) The places gazed upon are for purposes not directly connected with paid work and they normally offer some distinctive contrasts with work (both paid and unpaid).

(5) A substantial proportion of the population of modern societies engages in such tourist practices; new socialised forms of provision are developed in order to cope with the mass character of the gaze of tourists (as opposed to the individual character of 'travel').

(6) Places are chosen to be gazed upon because there is anticipation, especially through daydreaming and fantasy, of intense pleasures, either on a different scale or involving different senses from those customarily encountered. Such anticipation is constructed and sustained through a variety of non-tourist practices, such as film, TV, literature, magazines, records and videos, which construct and reinforce that gaze.

(7) The tourist gaze is directed to features of landscape and townscape which separate them off from everyday experience. Such aspects are viewed because they are taken to be in some sense out of the ordinary. The viewing of such tourist sights often involves different forms of social patterning, with a much greater sensitivity to visual elements of landscape or townscape than normally found in everyday life. People linger over such a gaze which is then normally visually objectified or captured through photographs, postcards, films, models and so on. These enable the gaze to be endlessly reproduced and recaptured.

(8) The gaze is constructed through signs, and tourism involves the collection of signs. When tourists see two people kissing in Paris what the capture in the gaze is 'timeless romantic Paris'. When a small village in England is seen, what they gaze upon is the 'real olde England'.

(9) An array of tourist professionals develop who attempt to reproduce ever new objects of the tourist gaze. These objects are located in a complex and changing hierarchy.

Look! Urry's description almost provides us an all-inclusive list about the constitution of tourist world. From the angle of space and time, there is a clear starting-point and an expectable end-point or returning-point for tourist world. The starting-point is the place and time where and when a tourist departs; and the returning-point is in the same place but a different time. The spatio-temporal continuum with a starting-point and returning-point constitutes the crust of tourist world. Attached to it, the tourist world shows its richness, uniqueness, independence and newness. It is the existence of this crust that creates an attractive tourist world different from the life-world. If we make a more generalized analysis of tourist phenomena, the basic elements and contents can be explicitly given as follows:

(1) Tourist space-time relation as the crust. Tourism is engaged in a certain space and time functioned as a restriction as well as a condition. Generally speaking, the space-time frame is characterized by temporariness and foreignness. As for tourists, he begins experiences and ends tourist activities within this space-time frame. As for destinations, all the impacts, all the involvement and all the forces actively or passively accepted from tourism are connected with this space-time frame, and some are directly occur in it.

(2) Tourist attraction system as external incentive. Tourism activity is meaningful and purposeful. Its meaning and purpose are determined by a tourist's (to be accurate, by a potential tourist) need and motive. Whereas the achievement of the meaning and purpose is dependent on the tourist attraction system functioned as external incentive to tourist activity. In this system, different elements serve tourists externally. Some are tourist objects which have not been developed by tourism industry; some are tourist resources; and some are tourist objects developed by tourism industry, such as tourist product[1]. These elements, which are relatively external to tourists, constitute fundamental tourist elements enclosed in the tourist space-time relation.

---

[1]   The viewpoint adopted here by the author is derived from the author's book, *Fundamentals of Tourism (second edition)* (China Travel and Tourism Press, 2004). See relative chapters in this book.

(3) Tourists as leading factors of tourist practice. Tourism is simply or firstly meaningful to the tourist. It can also be understood as meanings of tourism are given by tourists[1]. The occurrence of tourism must be the outcome actively pursued by potential tourists; therefore, tourists act as director, designer, performer and achiever of tourist activities. In the whole tourist world, tourists' needs and their satisfactions work as the basic forces restricting the interrelationship between the elements of the tourist world and directing the development and its pattern of tourist practice. Meanwhile, they also work as leading factors arising and eliminating different contradictions pertinent to tourism. Without tourists, there will be of no tourism, no tourist world, no tourist experience or tourist setting. From this point, tourists are leading actors in tourist world. The fundamental meaning or value of tourist world is to satisfy individual tourist's needs. But to what degree these needs can be satisfied is quite subjective.

(4) Interactions between tourists and others as relationships. Interactions between tourists and others form the most important humanistic phenomenon in tourist world. Some interactions will definitely occur due to the purpose of tourism, some will occur by chance in the process of tourism; and some belong to contractual trade relationship. All these interactions compose complicated tourist relationships including the interaction between tourists and hosts, between different tourists, and between tourists and tour operators. This network frames up tourist social network and at the same time, it creates different tourist settings. Therefore, these relationships will be a basic measure of tourist experience quality. In the process of interaction, tourist world will appear to be surprising and complicated. To understand the essence of the relationships, it is necessary to enter into tourist settings and discern tourists' motives and intentions.

(5) Intermediaries of tourism as a support. In the process of tourism, there are some intermediaries which facilitate tourist activities and contribute to formation of meaning and improvement of quality. They are infrastructure or hospitality establishments streamlining tourist process. In tourist practice of modern society, intermediaries typically refer to tourism industry and section as well as other relevant sections, such as traffic industry, communications industry, hospitality industry, shopping industry and travel agency industry. All these will be utilized by tourists to create and improve their tourist experience.

(6) Symbolic factors as the producer and transmitter of tourist meaning. In the process of tourism, as for tourist subject (tourists), they need some basic

---

[1]　Overlooking this viewpoint, a lot of people are trapped in the statement that tourist resources are movable. If they know a little about phenomenology, they will no longer feel confused.

symbolic instruments to help them obtain the meanings. With a symbolic nature, these instruments attach to tourist subject, tourist object and their intermediaries and create specific tourist settings, thus construct tourist world and establish meanings of tourism. Normally, these symbolic factors have certain symbolized meanings. Functioning as signifier, they construct tourist world as if they were setting up a stage for tourists to experience, displaying stage properties, controlling the pace of tourist activities, making tourists to be either spectators or actors. Tourist settings organized in this way have the attributes of stage or theatre. It is these symbolic factors that make a great contribution to the richness of tourist world and the composition of tourist experience. Actively confronting these symbolic factors and interpreting them in the right way is the key to obtaining the meaning of tourist experience.

After our deconstruction and analysis, we come to know that tourist world is a world with a complicated structure. People playing the leading role in this world are always elites in our society. Just as natural attitude and intersubjectivity exist in life-world, they also exist in tourist world. However, the functions of tourist world are quite different from those in life-world; hence the tourist world is only an escape from life-world. Meanwhile it has relatively independent meanings.

Judging from this model, we can see that the province and method of theoretical research on tourism can be expanded greatly. Not only tourists but also tourism researchers will be attracted by the richness of tourist world!

References

[1] Spiegelberg H.. *The Phenomenological Movement* [M] . Trans.Wang Bingwen, Zhang Jinyan. Beijing: Commercial Press, 1995. 216–217. (in Chinese). 赫伯特 · 施皮格伯格；王炳文：《现象学运动》[M]，张金言译，北京，商务印书馆，1995 年，第 216~217 页。

[2] Natanson M.A.. *Edmund Husserl: Philosopher of Infinite Tasks* [M] . Taibei: Taibei Yunchen Press, 1982. 159. (in Chinese). 那坦森：《现象学宗师：胡塞尔》[M]，台北允晨出版公司，1982 年，第 159 页。

[3] Schutz A.. *The Phenomenology of Social World* [M] . Northwestern University Press, 1967. 125–245.

[4] Lowenthal D.. Geography, Experience, and Imagination: Towards a Geographical Epistemology [M] . *Annals of the Association of American Geographers*, 1961. 241–260.

[5] Krippendorf J.. The new tourist: turning point for leisure and travel [J] . *Tourism Management*, 1986, (June).

[6] Urry J.. *The Tourist Gaze: Leisure and Travel in Contemporary Societies* [M] . SAGE Publications, 2000. 2–3.

Strategic Thinking of the Current Tourism Development

Wang Zhifa

(China National Tourism Administration, Beijing100740, China)

Abstract: China's tourism industry is under an important phase of upgrading and transformation. On a macro–strategy level, this paper tries to reflect and study on six items closely related to the building of a strong tourism nation: the staged characteristics and industrial functions of tourism industry, the mode of tourism economic growth and the promotion of industrial quality, the industrialization of tourism and upgrading of this industry, the furthering of opening–up and cultivation of a unified market, the increasing of tourism supply and stimulating tourism consumption, the protection of tourism resources and sustainable development. The paper is supposed to be of reference to the current study and decision–making on tourism.

Key words: strategic thinking; industrial quality; upgrading and transformation; furthering of opening–up

Since the adoption of the policy of reform and opening up, China's tourism has achieved a great success, and has made its due contributions to the development of China's national economy and society. The most remarkable success is that China has developed from a big country with rich tourism resources into a large tourism country in the world, and tourism has become a new growth point of the

---

[About author] This article is edited according to the speech delivered by Wang Zhifa, vice-chairman of the China National Tourism Administration, at the Provincial and Municipal Tourism Work Conference held not long ago. Slight increase and deletion have been made when it is published.

national economy instead of an important supplement to diplomatic work.

Since the 21st century, remarkable changes have taken place in the situation and tasks of China's tourism. In a word, tourism is now in the golden development period, the indusial transition period, the strategic upgrade period and the contradiction protruding period. Making China into a strong tourism country in the world as soon as possible and cultivating tourism into an important industry of the national economy, have become the glorious missions and historical tasks bestowed to the tourism industry by the era.

At present, tourism has many remarkable development opportunities. In the 11th Five-Year Plan period, or even longer, tourism will maintain a quick growth tendency. However, due to the restrictions of the system mechanisms accumulated over a long period of time, some deep-level contractions still exist while new problems keep appearing, resulting in the coexistence between development opportunities and the new and the old contradictions, which are intertwined with each other. Only if tourism implements the scientific viewpoint of development in an all-round way and realizes the strategic transition and strategic upgrade possible can it enter a higher development stage. Therefore, we need to pay particular attention to and consider the following issues in terms of the strategic research on the tourism development and macro decision-making.

1. Preiodic characteristics and industrial functions of tourism

The industrial quality, orientation and functions of tourism are the hot topics in the tourism academic circles, as well as a realistic issue that the government decision-making departments cannot avoid. The functions of any industry are not invariable, but must adapt to the country's economic, political and social systems. With reform and opening-up as the mark, the functions of tourism have been transformed from "institution" into "industry". Now the functions of tourism are changing quietly, i.e., gradually changing from "industry" to one with double characteristics of the economic industry and social functions. Now the functions of tourism have remarkably exceeded the economic category, and the contributions of tourism have broken through the single economic concept of increasing income and earning foreign exchanges. The comprehensive functions of tourism, rather than the single economic function, are becoming remarkable with each passing day in setting up the scientific viewpoint of development and the new resource viewpoint, that promoting the readjustment of the industrial structure and regional economic development; in speeding up the paces of the central and western regions to cast off poverty and become prosperous, and promoting the prosperity and stability in the frontier and ethnic minority areas; and in upgrading the people and masses' needs in spiritual and cultural life, promoting the building of material civilization and spiritual civilization and forming the socialist harmonious society.

At present, the functions of tourism have some remarkable characteristics as follows:

(1) The unification of the "people's livelihood" and the "national economy"

In 2007, the total number of overseas tourists arrival to China reached 125 million; the domestic tourist arrival exceeded 1.394 billion; the total income of tourism came to 893.5 billion yuan; and the number of tourism employees was only 10 million. Now tourism in China has become a large industry related to the livelihood of hundreds of millions of Chinese people. Directly perceived through the senses, it seems that tourism was not an important industry related to the national security and economic lifeline. However, tourism has affected the national economy and social development in many aspects, and influenced the country's overall "soft actual strength". A city with underdeveloped tourism can not realize internationalization, modernization and IT; and a country with depressed tourism can hardly be regarded as a prosperous country with good government and harmonious people.

(2) Paying equal attention to the economic function and social function

Though the income and foreign exchanges of tourism are increasing with each passing day, its economic function can no longer override anything. Its social function is becoming more and more obvious and important, and tourism is neither the appendant nor the derivative branch of the economic function. Now it can keep pace with the economic function, and moreover it has exceeded the economic function in the tourism developed areas. Therefore, when making an objective appraisal of tourism, we should prevent from overstressing the economic function, and avoid giving a general account of the functions of tourism. We should bring the economic function and the social function into the "main course" of work in the equal way.

(3) The exterior and the interior mutuality in Tourism activities and cultural attribute

Culture is the soul of tourism, and tourism is the carrier of culture. Without culture, tourism will lack charm, and without tourism, culture will lack vitality. Only if we realize the close combination between tourism and culture can we better promote material civilization and spiritual civilization.

(4) The special attributes of the multiple concepts and the opening-up of tourism

Tourism is a multiple and complicated concept. According to the general understanding, "tourism" means doing sightseeing and enjoying holidays and leisure. Obviously, it is only the "lesser" tourism concept. According to the traditional and imprecise understanding, tourism includes "six main elements," i.e., food, accommodation, transportation, sightseeing, shopping and recreation. It is the "medium" tourism concept. According to the international common definition of tourism, tourism should refer to that the "people leave their habitual living environment to go to some places and stay for a time for leisure, business or other purposes." It is the "greater" concept of tourism, which is much more extensive than the habitual understating of tourism, and it contains visit, sightseeing, holiday, leisure, recuperation, home leave and business trip. It also includes economic, S & T, cultural, educational and religious activities, and various new tourism tendencies that keeps appearing one after another. Therefore, developing tourism not only is the use of the landscapes, cultural relics, historical sites, and ethic customs and lifestyle, but also has a boundless development space.

2. Growth mode of the tourism economy and the improvement of the industrial quality

Upgrading the quality of tourism is the only way for making China into a strong tourism country in the world, and the key point is how to change the growth mode of the tourism economy. In 1990s, tourism put forward two "changes": The first is the change of the economic system, and the second is the change of the development mode. However, no remarkable progress has been made up to now. Quite a number of competitions among tourism enterprises rely on the "price war" instead of reducing cost and improving quality; the occupancy rate of hotels is not high; travel services operate with low profits or losses; tourist attractions make profits through increasing the admission ticket price; the competitions among tourism automobile companies are sharp; tourism products and service are fairly rough, and the means of sales promotion are unitary and backward; and the disparity of the reception benefits of inbound tourists between China and the countries with developed tourism in the world is remarkable. The key to the slow change in the growth mode of the tourism economy lies on that some regions and some enterprises have not fundamentally set up the scientific viewpoint of tourism development, nor fundamentally replaced the unilateral pursuit for the scale and speed with the improvement of quality and the increase of economic benefits, and nor fundamentally broken through the shackles of the traditional development ideology and development mode. It is the root of the problems and the focal points that China should solve in the future.

(1) Firmly setting up the scientific viewpoint of tourism development

Tourism in China started from scratch, and once summed lots of experiences in that era, such as the starting development of "building a gate to collect admission fees"; the development mode of "making exploitation and investment and getting benefits simultaneously"; and the "rolling development" through "cultivating tourism through tourism". Any experience is an outcome of the concrete environment and conditions, so that it does not have any wide adaptability in time and space. Accordingly, the scientific viewpoint of tourism development should keep pace with the times rather than standing still and refusing to make progress or following the same old rut for a long time. Today, when it is urgent to change the growth mode of tourism, innovative ideology, methods and experiences are especially required. The past experiences do not always adapt to today; and the experiences of large tourism countries do not always suit for the building of a strong tourism country. What we should explore and create is how to develop tourism according to high standards, high quality and high economic benefits. It looks like a tactical issue of the development method, but substantively, it is a strategic issue of the development viewpoint.

(2) Working hard to build a scientific tourism target system

We should take the formation of scientific development orientation as the target, and break through the thinking inertia of making appraisal only according to the scale, quantity and speed, and establish a target system that reflects the economic benefits, cost, quality and contributions. This system should include both the concepts of the tourism industrial scale and speed, and the targets for the comprehensive contribution rate and the market competitiveness, such as stimulating investments, arranging employment, increasing tax revenue, promoting environmental protection and optimizing ecology, and finally reaching the purposes of agglomerating and improving comprehensive competitiveness.

(3) Deeply expanding the development space of tourism

Satisfying the increasing tourism consumption demands, we should not rely only on exploiting new resources and building new facilities, but must make efforts in the deep use and sustainable development. Tourism consumption happens repeatedly; the consumption space is boundless, and the unexploited tourism resources are reducing day by day. Tourism consumption features a low consumption rate, repetition and multiple levels. Through the transformation and upgrade of the existing tourism products, we may keep expanding the market supply. The new development needs quite a large amount of investments and comprehensive matching, and the investment, use and return on investment need a certain period. Hence, it is impossible for tourism development to expand the scale without limitations, and intensive development is the inevitable choice for making China into a large tourism country.

3. Industrialization of tourism and the industrial upgrade

Industrialization is a work objective of an industry. To tourism, industrialization is the conditions and basis for realizing the upgrade of the tourism industry, and an inevitable choice for China to walk from "a large tourism country" to "a strong tourism country". The transformation of tourism from "institution" to "industry" is the basis for walking toward industrialization, but it does not equal to industrialization. The "six major elements" is the necessary fields that industrialization will involve, but is not equivalent to industrialization. Industrialization refers to that tourism operates completely and thoroughly according to the industrial rules, which is the requirement for making the operation mechanisms of the tourist industry become more perfect. Over the past three decades, we have always advocated to develop tourism as an industry, and tried every way to mobilize the enthusiasm of various aspects. A series of supporting policies have been promulgated, involving the positioning, policies and legislation of the tourism industry and the support by capital, which have played an important promoting role in the tourism development. However, these measures mainly focus on expanding the tourism scale and promoting the growth of tourism's main elements, and only a few involve the tourism industrialization system.

Realizing the industrialization of tourism includes the following contents:

(1) Promoting the expansion and fission of the key elements of tourism. We shall expand the development of the single industrial element into an industrial chain, and establish the close relationship with the relevant industries through stretching the industrial chain, so as to make it become the close-layer partner of tourism.

(2) Establishing the benign development of tourism. All the investors, developers and operators should follow the development law of tourism and establish the market revenue-sharing relationship with reasonable division of work. The grouplization, netlization and cooperative that conform to the market economic law should be realized between tourism enterprises on upper and lower reaches, between relevant enterprises and between enterprises of same type.

(3) Improving the regional competitiveness of tourism and the core competitiveness of enterprises.

(4) Forming a macro environment and tourism consumption environment favorable to tourism development.

(5) Expanding and increasing the comprehensive contribution rate.

As for speeding up the industrialized construction of tourism, the China National Tourism Administration put forward the demands in the five aspects at the National Tourism Work Conference in January 2006 on the basis of deep research, and formed comparatively systematic working thoughts.

(1) Founding a perfect tourism industrial market system. We shall make efforts to promote and make the regulation system, the building of tourism trade and the enterprise credit systems.

(2) Establishing the open and forgiving system of key element in tourism. We shall make efforts to promote the relevant social resources to gather around tourism industry.

(3) Establishing a perfect the tourism destination system. We shall put stress on promoting the building of the three grass-roots tourism destinations— tourism cities, counties and townships & towns.

(4) Further perfecting the system of tourism products. When upgrading the sightseeing products in an all-round way, we shall speed up the building of the leisure and holiday product system, and create the compound product system with Chinese characteristics.

(5) Building and forming the system of the talent and the tourism education that meet the market demands.

4. Expanding opening–up and cultivating a unified market

Tourism is an innate open industry. The unlimited tourism destinations, the diversity of the key elements of tourism consumption, the multi-level tourist sources and the integration of the tourism market have decided that tourism should be highly open. However, over a long period of time, due to the influences of local and trade interests, various types of protective fortresses have been formed at the tourism market, Such as foreign investors are restricted in the establishment of tourism enterprises; the development of networks by travel services of other places in the country is restricted; the operation of tourism vehicles of other places in the country is restricted; and guides of other places in the country are restricted in serving as local guides. These self-protection fortresses have seriously hindered the tourism market's opening up in the country. It influenced the full competition of the main body, and it become the deep-level obstacles of the industrialization, marketlization and internationalization of tourism, it led the lagging behind the domestic opening up objectively, and affected the bringing into full play the effects of opening to the outside world. Therefore, truly deepening and expanding domestic opening-up is the conditions and basis for further making tourism open wider to the outside world, and it is an imminent important work as well.

(1) Implementing the Administrative License Law in great earnest. We shall resolutely eliminate various types of protective fortresses, and prevent from the birth of new fortress-type obstacles.

(2) Speeding up the opening-up of travel services domestically. To honor commitments made during China's accession to the WTO, the China National Tourism Administration has promulgated in advance that the restrictions on "setting up branches" by foreign-funded travel services are cancelled. To promote the development and the progress through opening-up and, we must take the solution of the long-term problems of travel services as the main objective of opening up, and truly propel the professional division of work of travel services and netlization. To realize the above-mentioned objectives, we must stress on the pace of domestic opening-up which must be in harmony with that of the opening up to the outside world. We shall take the elimination of self-protection fortresses as the guidance and basis, and that the expansion of opening-up of the tourism market aims to upgrade the service of the tourism industry. Therefore, when studying the tourism market and the industrial supervision means, tourism departments should further emancipate the mind, and keep pace with the times, boldly do away with the system- and policy-related obstacles that restrict the development of tourism. It should lift controls over the permits for travel services to serve as operation agents and become members of an alliance, resolutely eliminate the self-protection regional fortresses and exclusive restrictions through using more legal and market means, and take the lead in founding the unified large tourism market in China and open the tourism market wider at appropriate time. Gradually we realize the integration of the domestic and foreign markets, and work hard for the maximum economic benefits of opening to the outside world.

(3) Speeding up the upgrade of tourism enterprises' competitiveness. In the early 1990s, the China National Tourism Administration advocated and initiated the system reform of all the tourism enterprises. At present, state-owned capital has basically withdrawn from the tourism enterprises in the provinces and

municipalities of East China, but the central and western China lag far behind. The reform system of large tourism enterprises is fairly slow, which has become an important cause why the tourism enterprises "small, scattered, weak and bad" have not become large and strong over a long period of time. During the 11th Five-Year Plan period, whether tourism enterprises can fundamentally improve the competitiveness at the market or not is an issue of mechanism-related significance, which depends on the system of enterprises, and the key way out is to speed up the reform paces of the enterprises.

5. Improving tourism supply and stimulating tourism consumption

Tourism consumption has the market characteristics of seeking new thoughts and making upgrade step by step. In the early 1990s, the tourism circles of China realized this market law and put forward the market concept of developing holiday and tourism products and perfecting the products' structure. At present, China has over 120 provincial-level tourism and holiday zones; a number of ecological, marine, rural, sports, health-building and yacht tourism products have emerged, and the supporting degree of tourism products has been remarkably improved. On the whole, the development of China's tourism products and market are in the primary stage: The tourism products take sightseeing as the mainstay, and the development thinking is mainly guided by sightseeing. The travel services mainly seek the reception of a large number of tourists. Now China's tourism has many rough tourism products, but a small number of quality products and it owns many resource-related tourism products, but a small number of cultural-upgraded ones. The supply of products can only satisfy the elementary and popular market, with insufficient personalization and comfortableness. At present, the supply level of tourism products is not high, and it has become an important element that restricts the upgrade of China's tourist industry.

(1) We should set up the development concept oriented by the market demand. The "resource-oriented" which means exploiting tourism products according to resources and then considering how to promote the products at the market obviously lags behind the needs of the development of the tourism market, and this viewpoint must be replaced by the "market-oriented" concept as soon as possible. Though the tourism demanding market undoubtedly keeps expanding, the expansion of the market contains the scale and the demand for the quality improvement as well. Only if we improve the product supply level

correspondingly and in advance can we further satisfy the market demand, promote the market consumption and stimulate the market supply. It is the characteristic of tourism and other service commodities, which is different from ordinary commodities, and an important content for the transition of the growth mode of the tourism economy.

(2) We should divide the tourism market in a scientific way. To satisfy the demand of the tourism market, tourism products must be classified and corresponded. There are many types and methods for classification. In a word, it is impossible for a high-end tourism product to involve itself with a large flow of reception, a low price and meager profits. The correspondence should be a long-distance tourism market, which can fully reflect the tourism, cultural and ecological value, and it embodies the tourism supplying service worthy for the price. The popularized tourism products will have a large flow of reception; the popularization should be reflected in the food, hotels, means of transportation and service; and the market classification should take the short-distance market as the mainstay. For instance, due to its unique tourism resources, fragile ecological environment, bad accessible facilities and high travel cost, a trip to Tibet should be defined as the high-end tourism product; the price for a tour of Tibet should be almost same as that for a trip to Europe; and the number of tourists to Tibet should be controlled appropriately. Only in this way, can we realize the coordination and harmony between the tourism development and the relevant demands. Since the Qinghai-Tibet Railway was open to traffic, the passenger transport capability has improved greatly, and the research on this issue has become very realistic and imminent. In spite of many different research conclusions, the importance of the products and the detailed classification of the market will be shown strategically.

(3) We should embody the culture and the characteristics. It is the key to improve tourism products, as well as the important competitiveness for winning the tourism resource market. The tourism products lacking culture can only be supplied to the low-end market for the time being, and it is impossible for them to have the long-lasting market competitiveness. Characteristics should be reflected in the tourism zones to avoid intensiveness and similarity. The hotels during the trip should also reflect the different grades to embody the varied purposes of the trip, holiday, business and meeting. In the design of the tourist itineraries, we should week through the old to bring forth the new, so as to better meet the demand of tourists of different levels.

(4) We should actively adapt ourselves to the tourism consumption law. At present, China's tourism market is being brought into line with the international

law at a high speed, i.e., China's tourism market is not suitable to the market demand, in terms of tourist groups, independent tourists, sightseeing, leisure, the development of products, popularization and sales promotion, resulting in the severe disjunction between the supply of products and consumption demand. Only if tourism enterprises initiatively correspond with the market demand and improve the products supply ability in an all-round way can they meet and stimulate the market demand.

6. Protection of tourism resources and the sustainable development

Tourism has always been regarded as the resource-conservation and environment-friendly industry. The possibility and potentiality are not equivalent to the inevitability and reality. In recent years, the destruction of the environment by the improper exploitation has often happened. It is worth noticing the elimination of tourism resources because of social and economic development. Therefore, to realize the sustainable development of tourism, we must fully realize the seriousness of the protection of tourism resources, and improve the recognition and protection sense of tourism resources in an all-round way.

(1) Adhering to the combination of strict protection and scientific exploitation

On the whole, the destruction of resources and environment because of improper tourism exploitation is a factor that affects the cultivation of a "resource-conservation" and "environment-friendly" industry, but it is not the most important factor. As the departments in charge of tourism development, in implementing the scientific viewpoint of development and promoting the sustainable development of tourism, the tourism bureaus at various levels should set up good examples, practice in an earnest way, act strictly according to the rules, stress the formulation of the plan, strengthen management, strictly make the evaluation of the environmental influence and the audit of the environment influence, resolutely correct various tourism exploitation actions that influence the resources and environment, advocate the ecological tourism consumption forms, guide tourists to strengthen the sense of environmental protection, work out the standards for products, service and management suitable for the sustainable development, form the green tourism management system, and promote the building of the tourism green industrial system.

(2) Improving the sense of the whole society in recognition and protection of tourism resources

Tourism resources are compound resources that are spread far and wide. No matter they are natural, cultural or social tourism resources; they coexist and integrate with other resources, such as hydropower, mineral, forest, wetland and river resources, which are more often than not tourism resources too. Therefore, it is very easy to cause conflicts with the protection and utilization of tourism resources during the exploitation and utilization of various resources. Such phenomenon is not rare in some regions. Though some economic benefits have been made, the tourism resources are subject to crashing destruction. The main reason that leads to this phenomenon is the people's insufficient understanding of tourism resources and the advantages of tourism, People cannot balance and unify the relationship between tourism and other industries, temporary interests and long-term interests, the economic interests and social interests in a comprehensive way, so let's protect tourism resources. As the investment in the resource-related exploitation has remarkably increased in recent years, various types of exploitation and construction have created unprecedented threats to tourism resources, with the hazardous degree much larger than the improper actions during the development and utilization of tourism resources. It is a fact that we should attract the high attention and vigilance of the governments at different levels and whole society.

(3) Strengthening coordination between the economic and social development and the protection of tourism resources

The improvement of the economic and social development level will be favorable to the improvement of the basis and conditions of tourism development and to the growth of the tourism market as well. If we cannot properly handle the relationship between development and the protection of resources, the destruction and extinction of tourism resources will be intensified. For instance, some regions adopted a series of unified actions, such as "eliminating thatched cottages," "making every village accessible by road" and "immigration relocation," so as to speeding up the casting off poverty and becoming prosperous. Under the promotion of special supplying capital and administrative forces, a large number of ancient villages, unusual civilian residences, ethnic and folk customs and culture, and the original ecological living environment were "urbanized," "modernized" and "standardized". Meanwhile, along with the improvement of the rural transportation conditions and the flourishing of the exchanges between urban and rural areas, the broad masses of peasants have initiatively changed the lifestyle and living conditions and the environment that have been handed down for thousands of years. It sought and modeled the new urbanized rural life, making the original ecological rural life that can be developed for sightseeing, folklore, rural and ecological tours undertake changes with each passing day. Accordingly a great number of tourism resources have fallen into oblivion and destruction. In a certain sense, the protection of tourism resources is the protection of the national culture and historical heritages and the protection, inheritance and development of spiritual civilization. Its profound significance is by no means of inferior to the development of material civilization.

References

［1］National Tourism Bureau. *China Tourism Yearbook in 2000–2005*［M］. Beijing: China tourism Press (in Chinese). 国家旅游局:《中国旅游年鉴》［M］，北京，中国旅游出版社，2000~2005。

［2］National Tourism Bureau. *China Statistical Tourism Yearbook in 2000–2005*［M］. Beijing: China tourism Press (in Chinese). 国家旅游局:《中国旅游统计年鉴》［M］，北京，中国旅游出版社，2000~2005。

［3］HE Guangwei. *To the Brilliance Times of China Tourism Industry*［M］. Beijing: China tourism Press, 2005(in Chinese). 何光昕:《走向辉煌的中国旅游业》［M］，北京，中国旅游出版社，2005。

［4］GAO Shunli. *Research on China Tourism Industry Policies*［M］. Beijing: China tourism Press, 2006(in Chinese). 高舜礼:《中国旅游产业政策研究》［M］，北京，中国旅游出版社，2006。

Several Important Issues Concerning China's Tourism Industry in the Transitional Period of Development

Ma Bo

(College of Tourism, Qingdao University, Qingdao 266071, China)

Abstract: Under the concerted actions of internal and external forces, China's tourism industry has entered a transitional period of development and it is confronted with profound changes in terms of its function, form, structure and driving force. There are six most important issues in the transitional period of development, namely: comprehensive control of tourism effect, internationalization, the building of marketization and public—private partnership, optimization of industrial structure and organization, integration of tourism and urban development as well as the upgrade of human resource development and scientific research level. The final solution of these issues depends on institutional innovation and its process is the process of institutional innovation of China's tourism industry.

Key words: China's tourism industry; transition; internationalization; marketization; institutional innovation

Introduction

Under the direct-drive of its own growth and the indirect-launch of the significant changes of external environment, the tourism industry in China has entered a

[About author] Mr. Ma Bo (1966-), Ph.D, dean and professor of College of Tourism of Qingdao University, Doctoral supervisor, with research interest centered on regional tourism development, tourism industrial economy and policy and tourism culturology. E-mail: mabo@qdu.edu.cn

transitional period of development. This judgment is becoming consensus of more and more scholars. In 1980s, Toffler has described the third wave like this, "Humanity faces a quantum leap forward. It faces the deepest social upheaval and creative restructuring of all time. Without clearly recognizing it, we are engaged in building a remarkable new civilization from the ground up. This is the meaning of the Third Wave."[1]Although compared with the third wave, the transition of China's tourism industry is a more specific topic ,and its complexity should not be underestimated. In fact, people "has not been well aware of it" so far. This article tries to macroscopically clear the transitional period of development of China's Tourism Industry, proposes and briefly analyzes six important issues with the purpose to consult experts and throw out a minnow to catch a whale.

1. The comprehensive control of tourism effect

In the last two decades of the 20th century, tourism industry in China basically lay under economic purpose. Although the academia had recommended Tourism Capacity theory and Host-guest Relationship theory from the west, it did not substantively influence the industry which pursues "development is the basic principle". After entering 21st century, inappropriate tourism development does increasing serious damage to recourses and environment, so the opposing voices to tourism are getting more, and the public gradually pay great attention to the environmental effects of tourism. In the last two years, with the promulgation of national policy on building a harmonious society, social and cultural effects of tourism also begin to be concerned. But by observing the current practice of tourism in many places, we can find that there are still great differences between local governments, investors and academia about the true comprehension of tourism effect. The economic effects of tourism are still placed in the supreme position, and the phenomena such as intrusion of nature reserve and urban water supply area, destruction of historical and cultural heritage, as well as serious neglect of community interests have occurred frequently.

The empirical studies on life cycle of tourism destination show that there are various main economic and social forces combined to determine the evolving process of tourism destination, such as environmentalists against the tourism industry or tourism development and other public, the forces of the government political and binding groups for or against tourism development. Based on the

---

①   Allin Toffler. The Third Wave [M].Beijing: SDX Joint Publishing Company, 1984.

conatus and the essential pursuit of consumer market in the recent years, it can be seen that the comprehensive control of various tourism effects is not only the need for ultimate solicitude, but also affects the stable development of economy in tourism destination. Furthermore it can be concluded that tourism industry in the future should inevitably follow the sustainable development way factually, and undertake more environmental ethics and social responsibilities.

We can confirm and agree the nice starting points of environmentalists, but we also have sufficient reason to oppose taking scenic area as "forbidden area". We can emblazon the pursuit of absolute fairness of humanists, but we have to admit the huge power of capital. We realize the capital has incomparable influence, but we also understand the power of capital must be constrained, or it will become a scourge to destroy our spiritual home. The tourism effects can be divided into several aspects, but the generation of them is an integrated process. As the primary problem of China's tourism industry in the transitional period of development, the function of transition of tourism industry from economic domination to comprehensive control of economy, environment and social culture means studies on tourism effects must be changed from unilateral study to multiple. When we consider economic, environmental or social cultural effect of tourism together, researchers must have the capability of multidisciplinary theoretical analysis. When studying comprehensive control of several effects, obliviously we can only rely on powerful team to solve problems. Thus, the responsibility of studying the comprehensive control of tourism effects is burdensome and there is a long road ahead. Unfortunately, the government does not pay enough attention and support to this beneficial attempt.

On one side, it is the wide spread of scientific concept of development and harmonious social thought; on the other side, it is frequent occurrence of inappropriate tourism development phenomena and hurly against tourism industry caused by the phenomena. It is noted that the comprehensive control and behavior regulation of tourism practical activities brook no delay, and it is impossible to wait for the slow maturation of related theoretical study. Therefore, according to the practice, study on the comprehensive control of tourism effect must make a breakthrough from the focus on techniques, for example, improvement of tourism environmental impact evaluation method, design of community-participated route, equilibrium of distribution relationship among stakeholders, preliminary exploration on comprehensive evaluation system ,and so on. In the course of view change and technological progress, rearrangement of institution will be initiated naturally. Institutional innovation will play a further role in solidifying concept and optimizing technique, thus fundamentally guarantee China's tourism industry in the transitional period of development.

2. The internationalization of tourism industry

China's tourism industry had been with the characteristics of internationalization at its initial stage. Tourism industry played a demonstrative role in the process of entering WTO. However, internationalization is still one of the most important issues in the transitional period of China's tourism industry.

The increase of China's outbound tourism is the direct reason for the internationalization. Although someone is worried about the sustainable growth of the outbound tourism market, and suggests restricting its development, but the result of logic analysis shows that nothing could slowdown the pattern evolvement of international travel from single flow to double flows, except the opening-up policy of china is changed. What left to us is just how to adapt the tendency[1,2].

The flourish of domestic travel impels China's tourism market to be transformed into dualistic structure from unitary structure, and the spring up of outbound tourism makes the market turn to ternary structure. The shock to the tourism industry of the market structure evolvement is huge and profound. In fact, for a long time, the dualistic industry structure resulted from the dualistic market structure has influenced overall benefit level of China's tourism industry: 12 national tourist vacation resorts are kept going by painstaking effort, the high-star hotels are falling into price war, and the travel traffic congestion has never be solved. All these are the true portraiture of this problem. So far, the dualistic structure problem hasn't be solved completely, and the new structural contradictions caused by ternary structure are emerging, such as the fierce competition between the hot domestic travel and outbound tourism, the upgrade and update of the travel attractions, the improvement of tourism infrastructure and public service, and the promotion of tourism enterprises' service level. All of these problems will be put on the schedule. Certainly, the development of outbound tourism will improve the integral quality of China's tourism industry, and finally it will solve the problem of tourism multiply economic structure. However, it is not only a long-term process, but also a process with more input and higher cost, which will cause new operating pressure to tourism industry.

In order to achieve the goal of the balance of international tourism payments, China's tourism enterprises will follow the outbound tourist flows, and take the road of transnational investment and management. Doubtlessly, this is a challenging task. Tourism overseas investment, including direct investment (establishing tourism enterprises) and indirect investment (Capital Management), requires the domestic tourism enterprises to get bigger and stronger, and there will be a great demand on the compound managerial personnels who grasp foreign language, understand foreign culture, are familiar with business process and know foreign laws and industrial commercial and duty system. If these difficulties can be overcome, the quality of China's tourism industry will be greatly improved, and China's tourism will merge into the global tourism economy system.

3. The building of marketization and public–private partnership

China's tourism industry has entered a transitional period of development and it is confronted with changes not only in terms of its form, function and structure, but also in terms of driving force.

In the framework of classical economics and new classical economics, government and market have been considered as two different economic adjustment modes. Market is the foundational and prior resource allocation means, and government is limited and constrained by the market, remedying market failure is the exclusive standard that measures the rationality of government's function [3]. It is certain that as any industry, market is the most fundamental power that promotes tourism industry. However, as a newly emerging industry, tourism industry obviously has its own particularity. Because of the comprehensiveness of tourism demand, broad tourism industry is factually the aggregate which is loosely composed of several industries with different properties. After tourism industry enters the stage of destination development, many public sectors are involve in the process of tourism development. Owing to the involvement, part of tourism products possess the characteristics of quasi-public goods. Meanwhile, tourism effects begin to transmit and extend from economic aspect to environmental, social cultural ones, government's participation in tourism also becomes inevitable. Thus, even in countries with high-developed market economy, government plays an irreplaceable role in tourism development.

It is more special when it comes to China's tourism practice. Reviewing the history of development of China's tourism, it can be seen that, on the one hand, thanks to Reformation and Opening-up Policy, China's tourism has made great development and progress; on the other hand, "government leading mode" is the basic mode of China's tourism industry in the long term. There are another three points besides the basic underground of "big government" which cause the contradiction between market leading and government leading as following: First, the transition from planned economy to market economy in China is gradual, the market subject is still in the course of development, its power is limited, and it can not take on the lead of tourism industry; Secondly, in order to improve the international purchasing power of China, the government endows tourism industry with capacity to earn foreign exchange, which makes tourism industry become subordinate sector rather than completely independent industrial sector; Thirdly, because of the ownership system, most tourism recourses and tourism products belong to the government, and the government becomes the management body.

"Government leading mode" makes great contribution to expanding the scale and promoting the position of China's tourism industry. The development of China's tourism industry creates a miracle that China has become a big tourism country in the world in only 20 years, but the critical voices become more and more in the progress of development. Tourism industry is one kind of demand-push industry, "government leading mode" is beneficial to solve the interim contradiction that demand exceeds supply, but when the supply reaches a certain scale, and the relationship between demand and supply reverses, disadvantages of the mode are fully exposed, such as low efficiency of resource allocation, low level of growth of enterprises and industrial benefit and so on. In fact, the view that in different period of the development of tourism industry, the focus of government's function is different, and it should be adjusted from "pioneer" to "coordinator", has been proposed long time ago [4]. In the 21st century, the step of China integrating into world open economic system has been speeded up obviously, the government's function is being adjusted continuously, the market subject is being strengthened constantly, the capability to earn foreign exchange of tourism industry has been "disrupted" by plenty of trade surplus and foreign exchange reserve. The basic trend of the international competition tourism industry has been changed by the springing up of outbound tourism. This series of major changes all support the view that government is still the pioneer in some areas where tourism industry just start, but "government leading mode" is not suitable for China's condition any more as a whole, and marketization is the main power of further development of China's tourism industry, and the inevitable choice to realize the leap from world big tourism country to world strong tourism country.

The direct result of marketization is the growth and mightiness of enterprises as economic subject. The expression forms of marketization can be grouped under product marketization and elements marketization. The former refers to the opening degree of financial, capital, labor force, land and technological market. In the process of marketization, generally the product market growth is the guide, but when product market develops to a certain extent, the promoting effect of economy will be more dependent on the growth of elements marketization [5]. After nearly 30 years' development, the product market of China's tourism industry has developed to a certain extent, but the growth of elements market has been quite delayed. It can be said that the product marketization degree of China's tourism industry still need to be improved, but the growth of elements marketization is the key issue for China's tourism industry in the transitional period of development.

Emphasizing the marketization of tourism industry dose not mean ignoring the government's function, it specifies that government and market should work independently. Only when the two forces cooperate with each other, tourism industry in China can realize its sound and rapid development. Foreign scholars usually consider tourism supply as an amalgam [6], including private products, pure public products and quasi-public products, so public-private cooperation is crucial. The author holds that Public-Private Partnership (abbr.PPP) is the chosen mode for China's tourism industry in the transitional period of development. PPP mode obviously refers to institutional arrangement must give consideration to social (public) and enterprise (private) interests. The mode requires that government standing for public interest and enterprises pursuing economic benefit should establish compact partnership, develop their respective advantages, and realize their double wins with the limitation of the relationship between responsibility, right and interest which is symmetric and made clear by government and enterprises.

The key point of establishing and managing the PPP mode lies in choosing appropriate enterprises and defining the public-private relationship between responsibility, right and interest. The PPP mode will have different contents for different projects. For example, franchise for national scenic spot, "tourism development company" mode (abbr. TDC) for tourism new development areas, "build—operate—transfer" mode (abbr. BOT) for infrastructure construction. In 2005, the author proposed the framework of PPP mode for Qingdao International Beer Festival which can be studied deeply by researches (see table 1).

Table 1   Framework of Qingdao International Beer Festival's PPP mode

	Government	Enterprises
Responsibility	· Distinguishing for-profit project or non-profit project · Bidding, signing the contract and authorizing · Infrastructure construction · Operating non-profit project · Destination marketing · Organizing, coordinating and guaranteeing	· Signing the contract and authorized operation · Product extending design and enlarging industrial chain·Secondary merchant · Market development · Achieve economic targets predetermined · Brand value extending, maintenance and appreciation
Right	Ownership and its manifestation	Relatively independent managerial authority with limitation of contacts
Benefit	Social welfare, promoting city tourism and business, city brand and image, profit sharing based on balance of payments	Profit manifestation based on input of production factors

4. The optimization of tourism industrial structure and organization

Tourism industrial structure refers to the technical and economic connections as well as connecting styles between every component of broad tourism industry. In terms of structuralism, there are internal relations between evolution of industrial structure and economic growth, high shift rate of industrial structure will cause high growth rate of economic aggregate, and vice versa [7]. Since the implementationn of the Reform and Opening-up Policy, the total amount of China's tourism industry is maintaining high growth, the tourism industrial structure is shifting all the time. But the changing speed of industrial structure is obviously slower than the total amount's, which causes many structural contradictions such as contradictions among development of tourism attractions, tourism transportation, travel agency and hospitality business, and the manifestations of contradictions are different in different area. The rationalization of composition of tourism industrial structure becomes one of the contents concerning China's Tourism Industry in the Transitional Period of Development, and it is also the technical force of growth in the transitional period of development.

Francois Vellas, a French scholar, pointed out that when tourism industry develops to a certain extent, because of large demand to infrastructure and accommodation facilities, it has been a capital-intensive industry as the heavy industry, but not the one with the characteristics of small investments and yield quick returns any more [8]. One of manifestations in the current transitional period of development is that the proportion in total elements of capital and technical elements gradually rises, and the proportion will be future improved by rising transnational development. The increase in the proportion will inevitably extend industrial chain of tourism, or rather require extension of industrial chain. Due to the fact that tourism industry belongs to service industry, the direction of extending only can be toward back. Although tourism industrial chain is used as a term by people, specific studies on extending it, such as direction, mode, extent, form of the chain and so on are little mentioned and needed urgently.

In terms of industrial organization, the most conspicuous problem of China's tourism industry is that the industrial concentration is relatively low, and tourism enterprises are small, separated, weak and lacking, which causes lowness of the whole economic performance and weakness of competitive ability of the industry. Since 1990s, great attention has been gradually paid to this problem, the step of tourism enterprises' grouping becomes faster and faster, and several tourism grouping enterprises with large-scale appear. However, unified market does not form due to the division between industry and regions, so the expansion of tourism enterprise grouping is actually quite difficult. Those tourism enterprise groups already set up mostly are the combination of hotels, hotels and travel agencies, hotels and tourism automobile companies, the merger and reorganization of vertical integration in strict terms appear then in the recent time. Grouping and transnational operation are in a reciprocal relationship and can develop together. Adequate capital, ample market network, advanced technology and management mode are the necessary conditions of tourism enterprises' transnational operation. And transnational operation also provides power and realistic condition for making full use of two kinds of resources and two markets. So, in the transitional period of development, tourism enterprise grouping will still be the focus of optimization of China's tourism industrial organization, and many types of market behavior, such as merger of enterprises, vertical integration and diversification and so on, will be the main stream.

Grouping of tourism enterprises is so fascinating that tourism SMEs are nearly neglected. SMEs can be covered up by big ones with economies of scale, and Marshall, the economist, quoted "story of grand forest" to explain this view [9]. In fact, SMEs are an important part of the modern economy. Guaranteeing the legal living right of SMEs is the precondition to maintain social stability and moderate market competition order in the process of economic revitalization. Tourism demand has the feature of fine crushing in area, volatility in time, personalization of tourists, and the characteristics of tourism product are non storage, simultaneity between production and consumption, etc. Thus, the organizational development of tourism industry can not be centralized as manufacturing industry. Even if in tourism developed countries, the SMEs occupy the absolute dominant in quantity. But they are so important not only because of the absolute advantage in the quantity, but also because they are the basic force of economic marketization and the important financial resources of local governments, and provide substantial employment opportunities. Facing the tide of grouping, the disadvantages on technological and managerial level, staff quality, production scale and capital accumulation of SMEs have been increasingly emerging, and many problems, such as low productivity, high cost, weak product development ability and market competitiveness, etc turn up. It shows that the government should concern and support tourism SMEs while encouraging the development of grouping. There are four kinds of methods to support them: Firstly, with the guidance to specialization and collaborative production, government should encourage and support the SMEs to participate in grouping strategy, develop multi-level and specialized facilities and co-ordination around large enterprises, and form specialized division system; Secondly, government should give necessary support with purpose of management technology and equipment modernization; Thirdly, government should encourage and support SMEs to establish cooperative organization in order to improve management level and decrease operating cost; Fourthly, government should give them financial tax preference.

5. The integration of tourism and urban development

Any economic activity will come to the reality. Tourism spatial distribution includes urban and rural. It goes without saying that urban is in the core position.

The prominent characteristic of urban tourism is the integration of tourism development and urban development. Urban tourism can only depend on the city and will inevitably be restricted by the urban planning, construction, and management. The integration of urban tourism resources needs the multi-sector cooperation. The development of urban tourism market needs the effective construction and communication of city image. The increase of urban tourism service level needs the improvement of urban public administration. Although cities are the main supporter of China's tourism industry and urban tourism receives much concern, the integration mechanism of tourism and urban development has never been formed. This problem has been commonly realized by academics, politics and industries in recent years. The management framework of urban tourism has been widely discussed and tried. The author considers that the adjustment of government tourism management organizations, establishing institutions with mightier functions, breaking department segmentation are all conductive to the integral development of tourism and city, but they are makeshifts at best. In order to solve the problem radically, adopting "small government, big market" and establishing the companies partnership mode mentioned above are very important. Certainly, government's scientific democratization and legalization construction are the premise and assurance to carry out the new mode.

Urbanization has always been the driving force of China's economic and social development. Joseph Stiglit, the Nobel Prize winner of economics, pointed out that China's urbanization is one of the two factors which effect the global society in 21st century, while it's also the greatest challenge of China in the new period. During the acceleration of urbanization, the driven force and mechanism of tourism development will be changed deeply. Meanwhile, tourism development is the important force for the urbanization development. Therefore, studies on the integration of tourism and urban development must be in the view of dynamic perspective.

Urban expansion shown as two modes in the whole world, one is single center structure while the other is multi-center network. At the beginning of the foreign cities development, it's the single center structure mostly, and then adjusted to the multi-center network after the mid-20th century. The main reasons are: 1) With upgrading of industrial structure, the proportion of tertiary industry and the "unurbanization" degree have been increased. 2) "Great urban disease" such as crowding, noisiness and high prices can't be bored by the people who require better life quality. 3)With the popularity of post- industrialization thoughts and the recognition of sustainable development, to standardize the evaluation of the cities turned out to be the humanity and environment but not economic. It means that the multi-center network mode is the choice of urban development. Urban planners divided the cities which expand scales fast since the Reform and Opening into two kinds. One is the cities maintain the single centre such as Beijing, Tianjin, Guangzhou, Jinan and Fuzhou; the other is the ones which explored new districts, including Shanghai, Shenzhen, Xiamen, Suzhou, Qingdao and Zhongshan. Research shows that almost all the cities with new districts develop better than the single center cities. The key factor is that new districts add effective supply to urban space by expanding with low cost[10]. The urban are both the production activities centers and the life activities centers. Urban planning and construction need to consider its economic function, as well as the social culture and environment function. We deduce that the multi-center cities and metropolitan area delimitation will lead to the way of China's future urbanization. The main trend will change the internal structure of urban tourism. In other words, the development of urban tourism should adjust to the changes in urban forms positively. It is known to all that China's economic industry is in the process of service oriented, on which the construction of satellite city and new district depend. Tourism is not only the important component of service industry but also the leading industry in developing service industry. Therefore, developing tourism is the essential content of urbanization. In conclusion, from the dynamic prospective, the integration of tourism and urban development has far-reaching significance.

In the process of metropolitan area delimitation, the structure relationship of center and its hinterland between cities will be strengthened. In fact, the structure has been manifested in Pearl River Delta, Yangtze River Delta, Beijing-Tianjin Area and the Area of Shandong Peninsula. In the recent 3 years, Rizhao, Weifang and Haiyang have proposed successively the strategy of developing tourism in line with Qingdao, which gradually established Qingdao as the tourism central city. In order to seek for the regional dividend, the tourism central city should aggregate the tourism industry elements as quickly as possible, form the "brain" of regional tourism, in order to play its radiating function. How to promote tourism elements market in tourism central cities? How does the city cooperate with the development of elements market? Obviously, related research and practice need to be strengthened.

6. Human resource development and scientific research

In fact, the low price of labor is one reason of the rapid development of China's tourism industry in the past 30 years. However, the present situation is changing. Compared with the historical change of compensation level in hospitality, we can find the compensation level of middle and high rank manager keeps rising, but the salary of ordinary staff did not change much, in some area even a little decreased. The former showed the lack of manager resources and the latter proved the rich labor resource in China. But recently there is a universal pressure of recruitment in hospitality, even appeared labor shortage. Both indicated that the labor price is so low. The supply structure of labor in hospitality is tending to balance or shortage from surplus, which will lead to increase of the cost of labor resources. In fact, some hotels have greatly raised staff salary. Undoubtedly, this variation and its impact to the transition of tourism industry should be paid much attention by authorities.

Developing tourism education is one effective ways to solve the human resource problem, including the lack of the managers and basic level staff. Since the Reform and Opening-up Policy, the tourism education in China developed rapidly, but was not closely connected to the industry. There is a significant disjunction between personnel education and trade practice. The author holds that reasons for this phenomenon are various. Except for the self-problem of tourism education, it is also closely related to the ignoring of human resource development by government and enterprises. The raising of labor cost, industry structure form's changing, as well as the fiercer competition among enterprises foreshow that the integration development of industry, teaching and research should lead by the enterprises instead of the university. Then a sound mechanism of tourism education will be built. Meanwhile, the structural optimization of tourism education must be transformed from only pursuit of upgrading school-running level to the balanced and realistic development of multiple levels.

There are a series of important issues in the transitional period of development of China's tourism industry. The practice has high expectation to the theory, but the supports and contributions of academia to the industries are clearly not satisfied. Someone thought the reason for that is the academia are not familiar with the industries [11]. Because of unacquaintance with industries, achievements made by researchers are divorced from reality, and even absurd. Studies on tourism can not be done in the classroom, but field research is only the first step of study. Tourism academia does not make great contribution to the industries, not only because it is not familiar with the industries, but also because of internal cause that the theories of it are not perfect. The ultimate aim of tourism studies is constructing the system and method that can describe, explain, predict tourism phenomenon scientificly. It can be imagined, if people speak strong dialect not Mandarin, communication will be inefficient and even confused, and if there is only chessmen without chessboard, and playing rule, chessmen will be useless. If there is no rigorous concept system and scientific paradigm, accurate observation, analysis and prediction on complex matter will be impossible. No matter in China or foreign countries, the core issue of tourism studies are abstract analysis, discovering general truth through special contradiction, and constructing theoretical system that can improve people's capability of understanding tourism, and lead people into exploring concrete tourism problem, but not describing and analyzing concrete tourism phenomena.

If the tourism studies in China reach maturity, it must go through the process of theory development. This process started when we treated the tourism study as a subject instead of a field. There are two ways in the process: First, abstract and generalize the practical experience by induction, then form the theory structure; Second, learn theories broadly from economics, management, sociology, anthropology, and resources & environmental science, etc, integrate and reform the theories and methods of many subjects creatively according to the special nature of tourism, finally form tourism theory frame. But what should be pointed out is that some people in tourism academia have a wrong thought that the theory and the practice are two simply opposite parts, and despise the theory. However, actually, although theory is one kind of abstract and systematic cognition, it can only arise from practice and experience. The aims of theory are explaining the experience and reality. "The Science Circle", raised by an American sociologist named Wallance, which is also widely used in social science, tells us that observing and theory are the two starting points of research. Experience world and theory world are always interacting and rising in a spiral [12].

"We do not appreciate those sociologists who do not observe but only talking about nothing, or those who do not think but only observing, or those who do not make their thinking tested by systematic experience but only thinking, no matter what authority they are" [13] Merton, Robert King, an America sociologist said. Studies on China's tourism just start, so real researchers anywhere must carefully understand the profound meaning of the saying.

The above discussion on China's tourism industry in the transitional period of development emphasized posing problems rather than solving problems. Even though, we can see that the transition development of China's tourism industry is a great subject, it's also a gradual and challenging process. The core of transition development is institutional innovation, and the final solution of these issues depends on institutional innovation and its process is the process of institutional innovation of China's tourism industry.

References

[1] Zhang Lingyun, Yang Chen. From "Giving Prominence to Foreign Exchange Earnings" to Balance of Payment: Rethinking about the Developing Strategy of China's Outbound Tourism [J] . *Tourism Tribune*, 2007(6): 20–24(in Chinese). 张凌云、杨晨:《从创汇优先到平衡收支:我国出境旅游发展战略的再认识》[J],《旅游学刊》2007 年第 6 期，第 20~24 页。

［2］Ma Bo, Kou Min. *An Initial Study on the Development of China's Outbound Travel and Its Impact*［J］. Tourism Tribune, 2006(7): 24–28(in Chinese). 马波、寇敏:《中国出境旅游发展及其影响的初步研究》［J］,《旅游学刊》2006 年第 7 期, 第 24~28 页。

［3］Huang Heng–Xue. *Public Economics*［M］. Bejing: Beijing University Press, 2002. 25–53(in Chinese). 黄恒学:《公共经济学》［M］, 北京, 北京大学出版社, 2002, 第 25~53 页。

［4］Kuang Lin. *Centralization or Decentralization: The Governments Are in A Dilemma in Developing Tourism Industry*［J］. Tourism Tribune, 2001(2): 23–26(in Chinese). 匡林:《集权还是分权 : 政府发展旅游业的两难境地》［J］,《旅游学刊》2001 年第 2 期, 第 23~26 页。

［5］Wang Xiaolu. *Marketization Process of China*［N］. China Economic Times, 2003–03–20(in Chinese). 王小鲁:《中国的市场化进程》［N］, 2003 年 3 月 20 日《中国经济时报》。

［6］Chris Cooper, John F, David G, Stephen W. *Tourism principles & practice*［M］. UK: Pitman. 1993. 80.

［7］Su Dongshui. *Industrial Economics*［M］. Beijing: Higher Education Press, 2000, 234(in Chinese). 苏东水:《产业经济学》［M］, 北京, 高等教育出版社, 2000, 第 234 页。

［8］Francois Vellas. *International Tourism Economic and Policy*［M］. Beijing: Tourism Education Press, 1989. 30(in Chinese). 弗朗索瓦 · 韦拉:《国际旅游经济与政策》［M］, 北京, 旅游教育出版社, 1989, 第 30 页。

［9］A Marshall. *Principles of Economics*［M］. Beijing: Commercial Press, 1983.17(in Chinese).A. 马歇尔:《经济学原理 (下册)》［M］, 北京, 商务印书馆, 1983, 第 17 页。

［10］Zhao Yanqing. Olympic Economy and Beijing spatial structure adjustment［J］. *City Planning Review*, 2002(8): 29–37(in Chinese). 赵燕菁:《奥运会经济与北京空间结构调整》［J］,《城市规划》2002 年第 8 期, 第 29~37 页。

［11］Wang JianMin. Mature studies on tourism should reflect familiarity with industry situation［J］. *Tourism Tribune*, 2007(1): 7–8(in Chinese). 王健民:《旅游研究的成熟要体现出学界对业界的熟悉》［J］,《旅游学刊》2007 年第 1 期, 第 7~8 页。

［12］Walter L Wallance. *The logic of science in sociology*［M］. Aldine. Atheon inc. 1971, 18.

［13］Robert Merton. *On theoretical sociology*［M］. Beijing: Huaxia Publishing House, 1990. 92(in Chinese). 罗伯特 · 默顿:《论理论社会学》［M］, 北京, 华夏出版社, 1990, 第 92 页。

A Research on Quantitative Evaluation of Tourism Growth Quality in China

Chen Xiuqiong[1,2], Huang Fucai[1]

(1. School of Management, Xiamen University, Xiamen 361005, China; 2. School of Management, Putian University, Putian 351100, China)

Abstract: The issue of industry growth quality is a hot topic in China's tourism industry with the problem of accumulation tourists while not benefiting efficiency of production factors. This paper discusses this topic by evaluating the growth quality of Chinese tourism industry from five aspects: product quality, environment quality, factors quality, developing model and tourism operation quality. The result shows that the comprehensive quality of tourism growth has witnessed a descending trend. The quality improvement and quantity expansion are adverse, and the growth quality of Chinese tourism industry is worrisome.

Key words: industry growth quality; tourism industry quantity; evaluation; China

Introduction

The concept of the growth quality of China's tourism industry is by no means something new. The growth quality of China's tourism industry has received an increased amount of attention in the last decade, especially with the serious

[About authors] Ms. Chen xiuqiong(1976-), Ph.D, lecturer of tourism management school of Putian University; Mr. Huang Fucai(1947-), professor of tourism in the school management at Xiamen University, with research interests centered on issues related to the theoretical studies of tourism.

problem of accumulation tourists while not benefiting efficiency of production factors. However, traditionally its research has been mainly focused on qualitative analysis; few researches are from quantitative evaluation.

To test China's tourism industry growth quality, this article is to examine some growth quality related issues with a quantitative perspective while rethinking of China's tourism industry quality and efficiency.

1. Method for measuring industry growth quality

In term of economic theory, economic growth is the calculation and the improvement or increase in value of production factors[1]. The calculation of production factors means the increase in both capital and labor, which is the source of quantity expansion of economic growth. While the improvement or increase in value of production factor refers to more efficiency in both capital and labor, which is the source of quality growth of economic development. Thus it is one way to enhance international competitiveness.

In term of its rapid development, tourism industry has a large market share in the world and has played an important role in the world. However another important thing is its growth quality and efficiency. Traditionally, we often evaluate industry growth quality from these five aspects:

Product quality

Product quality is an important index to test resource efficiency, which is the prerequisite and foundation of industry growth quality. On one hand, qualified product has its value and usage, which will prevent wasting of resource. Inversely, it will cause the wasting of resource. On the other hand, the quality of product will influence its competitiveness. Because the competitiveness has turned from price competent to quality competent. The product with high quality has more competitiveness. Thus the international competitiveness is decided by its product quality.

Product quality is often measured by product percentage of pass. However, it is critical to acknowledge the difference that exit between tourism industry and other industries. Tourism product has another characteristic-perishability. It relates to the fact that tourism products are intended to be consumed as they are produced. Because the consumption of the service and its production occur simultaneously, there is no inventory. This puts pressure on hospitality businesses to operate at as high a level of capacity as possible. Except for product percentage of pass, utilization efficiency is also critical for tourism product quality. Thus the indexes for tourism product quality include:

1. Satisfaction index

As discussed earlier, the "product" of tourism is different in nature from other products. The product of tourism is "a satisfying experience at a desired destination" (Jefferson & Lickorish, 1988:4) and "a bundle of benefits" (Heath & Wall, 1992:129) [2], and is intangible, inseparable, perishable, and heterogeneous. Furthermore, "the tourist travels to the destination area where the product is experienced" (Heath & Wall, 1992:126) [3].

The above discussion highlights the importance of tourist satisfaction. So we can say if tourist is satisfied with its product, which means tourism product is qualified. So we use satisfaction index for tourism product percentage of pass. Index IS stands for inbound tourists satisfaction and index DS stands for domestic tourist satisfaction.

2. Utilization index

On one hand, in Song's research [4], he refers that the boundaries of tourism industry are fairly diffuse and therefore it is hard to define its customer. Tourism industry is an industry cluster. It combines segments from other industries consisting of a wide variety of products and service. "There is probably no other industry with such diversity in terms of the sectors it embraces, such as restaurants, hotels, airlines, travel agencies, shops and so on"(Sara Nordin, 2003:15) [5]. On the other hand, most tourism products serve both for tourists as well as for residents and non-residents. In addition, it doesn't reach a common acknowledgement about the capacity of most tourism product. Thus it is difficult to measure the whole utilization efficiency of tourism in practice.

However, according to the statistics in 2003 in yearbook of China tourism statistics, the major investment on tourism industry focus on hotel sectors, which is 83% of investment on tourism industry. Thus room occupancy rate of star-level hotels (HO) can stand for utilization efficiency in tourism industry.

Environment quality

Environment is critical for industry growth quality. The future of tourism is inextricably linked to the quality of environment. Degrading the very environment on which tourism depends could undermine its business in the long term. With failure to recognize the importance of environment quality, development of tourism industry will lead to environmental degradation and worsen as well as over-exploitation, which will lose its development meaning. In addition, tourism industry depends to a large extent on environmental qualities than other industry. Tourism needs balmy weather, corals and coastlines-all depends on an excellent environment. If one area worsens its environment, it will lead to the end of tourism industry. Without qualified environment, in particular without clean environment, there would be no tourism activity. The Sustainable tourism heavily depends on a clean environment. However, environment is a kind of common-property resource and it serves for both tourism industry and non-tourism industry. Thus environment quality is the result of accumulative effects both from tourism industry and non-tourism industry. We can't give a reasonable appraisal on its quality just from one industry.

Production factor quality

According to classical economic theory on competitiveness advantages of an industry, production factor quality is the essence for industry growth quality, especial for its efficiency and quality development. In term of output, the more output the higher efficiency the better quality. Input factors of tourism industry mainly include man, power and capital. However, the core power of tourism industry is tourism attraction, which is confused to evaluate its value whether natural resource or human resource. So we just evaluate its capital and man resource efficiency.

1. Labor productivity index

Labor productivity index reflects the average value that every tourism employment creates, which is an important index for measuring the quality of labor factor.

If labor productivity increases every year, which means improvement of industry growth quality. The formula is following:

$$TRL_t = TY_t / TL_t \tag{1}$$

TRL_t refers to tourism labor productivity in t year. TY_t refers to tourism revenue in t year. TL_t refers to tourism employment number in t year.

2. Capital productivity index

Capital productivity index reflects the output of unit capital, which is an important index for factor quality. The higher the capital productivity the more output one unit inputs. The formula is following:

$$TRK_t = TY_t / TK_t \tag{2}$$

TRK_t refers to tourism capital productivity in t year. TY_t refers to tourism revenue in t year. TK_t refers to tourism investment in t year.

3. Incremental capital–output ratio (ICOR)

Incremental capital-output ratio is the rate between capital and incremental yield, which is an important index for efficiency of capital. The lower the ICOR the more efficiency the capital has.

$$TICOR_t=(TK_t-TK_{t-1})/(TY_t-TY_{t-1}) \tag{3}$$

$TICOR_t$ refers to tourism incremental capital-output ratio in t year. TY_t refers to tourism revenue in t year. TK_t refers to tourism investment in t year.

Way of industry growth

Way of industry growth or developing model is important for industry growth quality because it is the foundation of increasing efficiency of resource utilization. The input elasticity is rate between input growth and yielding growth of production factor. The input elasticity rate is an important index for judging the way of economic growth whether extensive mode or intensive mode, which is measuring efficiency of production factor. The input elasticity rate is an index for judging economic developing whether extensive mode or intensive mode. If the input elasticity rate is 80%, which means the contribution of growth rate of factor input had an absolute superiority of 80%. The high input elasticity rate implies that economic growth is high input, high consumption, high pollution, and low efficiency. Thus it is unsustainable.

1. Tourism labor input elasticity ratio

Labor input elasticity ratio reflects the extent that industry depends on growth rate of labor input. Those industries with larger input elasticity of labor should be more labor intensive. Formula of tourism labor input elasticity ratio is following:

$$TTL_t=TBL_t/TBY_t \tag{4}$$

TTL_t refers to tourism labor input elasticity ratio in t year. TBL_t refers to the growth rate of tourism direct employment in t year. TBY_t refers to the growth rate of tourism revenue in t year.

2. Tourism capital input elasticity ratio

Tourism capital input elasticity ratio reflects the extent that tourism industry depends on growth rate of capital input. Those industries with larger input elasticity of capital should be more capital intensive. Formula of tourism capital input elasticity ratio is the following:

$$TTK_t=TBK_t/TBY_t \tag{5}$$

TTL_t refers to tourism capital input elasticity ratio in t year. TBL_t refers to the growth rate of tourism investment in t year. TBY_t refers to the growth rate of tourism revenue in t year.

Industry operating quality

Industry operating quality is an important part of industry growth quality. Industry operating quality includes the operation of industry and industry structure. On one hand, smooth, stable and coordinate development of one industry will promote its sustainable development. However, if one industry fluctuates drastically, it will lead to the potential risk of resource waste and efficiency reduction. On the other hand, industry structure plays an important role in the efficiency of resource utilization.

Because of variable tourist needs, its consumption of the service and its production occur simultaneously, tourism industry growth quality should pay a special attention to its operating quality. Indexes for tourism industry operating quality include:

1. Industry development fluctuation ratio
Industry development fluctuation ratio is an index for measuring the change of industry growth rate. The formula is the following:

$$TRB_t=(TBY_t-TBY_{t-1})/TBY_{t-1} \tag{6}$$

TRB_t refers to the fluctuation ratio of tourism industry growth. TBY_t refers to the growth rate of tourism revenue in t year.

2. Industry structure
Industry structure is an important index for measuring efficiency of resources allocation. Reasonable industry structure implies that resources allocation is efficiency and helps to its sustainable development. The unreasonable industrial structure implies misallocation of resource between the different sectors, which may lead to waste of resource and contribute to unstable development.

Industry structure can be measured both from macro and micro term. In term of macro economy, the development of tourism industry should be compatible with economic development. However, it is still a controversial topic about what proportion tourism industry should take in national economy. So we may ignore this topic. In term of micro economy, different sectors of tourism industry should be compatible, well-shared and with balance of supply and need. Thus, tourism industry structure can be measured from the following:

$$C_t=D_t-S_t \tag{7}$$

C_t refers to the coordination rate of tourism industry in t year, D_t refers to the income elasticity of tourism industry demand in t year, S_t refers to the income elasticity of tourism industry supply in t year. If C=0, it implies that the supply equals to demand, which results in an economic equilibrium. If C>0, it means that below equilibrium, consumers demand more of the good than that of producers are prepared to supply. If C<0, it means that above equilibrium, consumers demand less of the good than that of producers are prepared to supply. Whether C>0 or C<0, it is an implication of no equilibrium.

In a word, we can conclude that industry growth quality is an integrated efficiency of these above indexes. Every index can affect industry quality. If we want to improve industry growth quality, we should pay attention to all these indexes.

2. Empirical study on china's tourism industry growth quality

Because of shortage of statistics and the unusual 2003 with SARS, this paper gives an empirical study on Chinese tourism industry growth quality from 1994 to 2002. The main sources of data used in this study are the series statistic annual books of Chinese tourism industry from 1994 to 2002.

Product quality

According to the statistics, except for the year of 1995, tourist satisfaction index is in the trend of increase, which implies the improvement of tourism service and competitiveness enhance of Chinese tourism industry. This is the result of the standardization and law-making effort of Chinese tourism bureau, which leads to improvement of tourism service, operation and management.

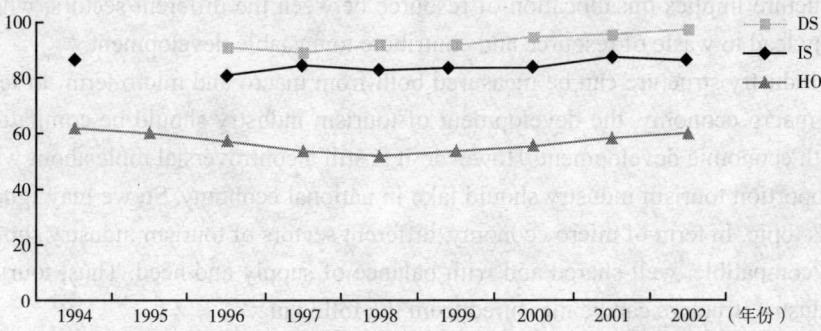

Figure 1   Product Quality of Chinese Tourism Industry, 1994-2002

In addition, the room occupancy rate of star-level hotels (HO) still remains in the low level, which implies over production and low efficiency of tourism industry. Except for 1994, 1995 and 2002 reaching Break Even Point-60%, all the other years the room occupancy rate of star-level hotels (HO) are lower than break even point. From 1994 to 1998 the room occupancy rate of star-level hotels (HO) was in the trend of decreasing while it was in the trend of increasing from 1999 to 2002, which means that efficiency of hotel industry got improvement.

Production factor quality

On the whole, production factor quality is in the trend of decreasing. Labor productivity is in increase from 1994-1997 while decrease sharply from 1997 to 1999. Labor productivity has a little increase from 2000-2002, which means China's tourism industry has a little improvement in labor productivity.

On the whole, China's tourism production factor quality has become worse since 1994 while got a little improved from 2000 to 2002. The decrease of production factor quality implies that the growth rate of China's tourism revenue is lower than that of received tourist number, which leads to decrease of profit ability and gives rise to a phenomenon of accumulating tourists while without benefiting wealth. The reasons are the following:

Firstly, on one hand, the global economic slowdown may lead to reduce the stay-length and expenditure of international tourist while keeping visit number. International tourists may choose cheaper hotel, flight and travel destination to reduce its expenditure. On the other hand, because of terrorism activity and many uncertainties, tourists may have short-haul travel instead of long-haul

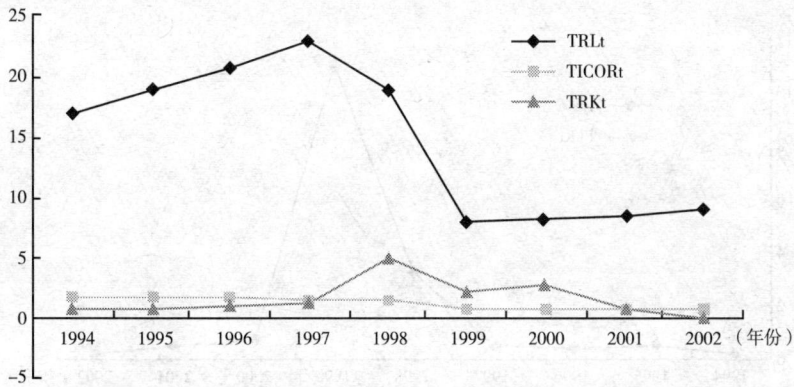

Figure 2   Factor Quality of Chinese Tourism Industry, 1994-2002

travel, which may lead to reducing the average expenditure of international tourists in every travel activity.

Secondly, it is caused by the sharp development of national tourism. However, on one hand, national tourism still remains on the phrase of lower expenditure, short-distance, and few stay-over, which lead to low productivity. On the other hand, visit numbers of national tourist keep increase while its short-haul travel is more frequent than long-haul travel. In despise of the increase of national tourist numbers and more expenditure for every travel, short-haul travel is much more than long-haul travel, which result in the growth rate of national tourist revenue lower than that of received tourist number. In addition, the major national tourists is from rural area, among which is 60%. The low ability of disposable income results in the low level of Chinese national tourism expenditure. Zhang's research [6] points out that average Chinese national tourism expenditure will increase unless rural disposable income gets improvement.

Thirdly, rough competition of China's tourism industry results in price reducing. China's tourism industry still focuses on price competition strategy. Nowadays, price reduction is the usual practice in tourism industry, especial in transportation sector and airline companies. However, competition in this way may lead to a quality decline and restrict its development, which results in low factor productivity and efficiency.

Way of industry growth

On the whole, the growth way of China's tourism industry remains on intensive mode. The growth way got intensive with more labor and capital input from 1994-1999. Especially in 1999, China's tourism industry was one of the highest consumption. However, China's tourism industry gradually got improvement

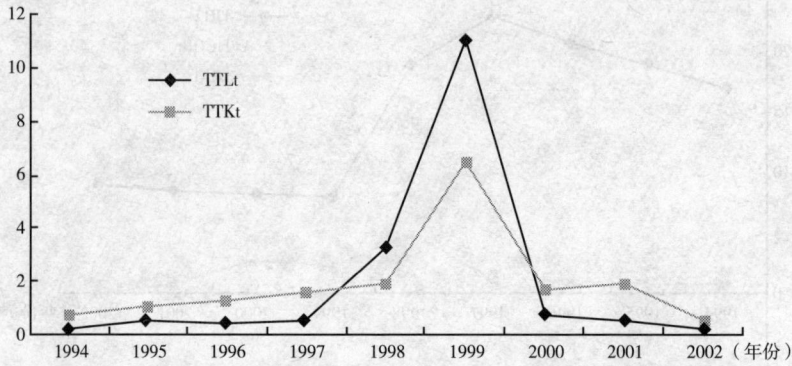

Figure 3   Quality of Growth Way of China's Tourism Industry, 1994-2002

with less input from 2000 to 2002.

Compared with national labor input elasticity ratio (TL), tourism industry takes advantages in absorbing employment. Chinese tourism labor input elasticity ratio (TTL) was 0.24-11 while Liu research [7] shows that Chinese national labor input elasticity ratio (TL) was 0.07-0.17. Except for 3.30 in 1998 and 11.02 in 1999, TTL remains 0.24-1. However, TTL reduces to 0.22 from 2000 to 2002, which implies decreasing on employment contribution of China's tourism industry as well as increasing on efficiency of labor input and better of employment quality.

Compared with national capital input elasticity ratio (TK), Chinese tourism is an high investment industry with more TTK. This implies that tourism industry highly depends on capital investment, which conflicts with the concept that tourism is a low investment industry. Chinese tourism capital input elasticity ratio (TTK) increased from 1994 to 1999, which means that capital efficiency decreased. However, Chinese tourism capital input elasticity ratio (TTK) got improved with TTK decreasing from 2000 to 2002. On the whole, growth way of Chinese tourism industry hasn't got improved these years. It remains in intensive mode with high consumption and high investment, which results in low efficiency of factor productivity. This may due to the following:

Firstly, government-dominated institution leads to low efficiency of production factors. Since major tourism resource is common-property, government-dominated model in tourism industry is prominent. For example, hotel sector takes up 80% investment of Chinese tourism industry. However, just as we all know, hotel sector is the most open to foreign invest in tourism industry, among which 79.85% was collective or state-owned in 1994 and 67% in 2002. What's more, the collective or state-owned hotels reached 72.6% in Beijing in 2002. However, all indexes and indicators of the collective or state-owned hotels were the worse in Beijing from 1997 to 2000 [8]. In addition, their profit ability decreased every year, which results in low efficiency of investment.

Secondly, tourism industry doesn't develop well. Due to repeated construction, low product quality, and low level of development and planning, the serious problem of "the assimilation phenomenon" result in rough competition for market share and waste of resource.

Industry operating quality

Due to some unfavorable events and factors home and abroad, China's tourism industry fluctuates obviously. With suddenly popular and then unpopular, Chinese tourism industry didn't keep steady development with 3 highlights in

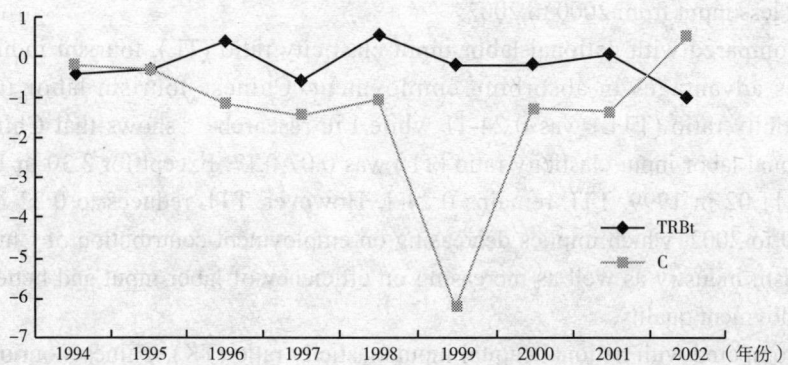

Figure 4   Operating Quality of China's Tourism Industry, 1994-2002

10 years. It is a contrast between frequent fluctuations and sharply expanding in tourism investment, which implies the low steady of input efficiency and high risk of investment.

Due to the concept of low investment and high output of tourism industry as well as the bettering of industry position, it has ever come out heated investment in Chinese tourism industry since 1990s. Thus it causes rapid growth of the supply and helps to settle some bottle problem of tourism industry. However, on one hand, over-heated investment caused many problems, such as repeated construction and over-development, which lead to low efficiency of resource. On the other hand, over-heated investment leads to no equilibrium of supply and demand and unreasonable industry structure. Except for supply lag for demand with C=0.54 in 2002, all the other C are below 0 and decrease all year round from 1994-2001, which implies that supply is more than demand in tourism industry. Particular in 1999, the value C reached to -6.25, which means that supply is much more than demand. On one hand, the over-heated investment leads to over-production and waste of resource. On the other hand, the over-heated investment leads to competition heat up and low profit, which reduce its efficiency of investment. Both hotels and travel agencies have come into low marginal profit period and its investment profit ability reduces sharply. Since the middle 1990s, profit ability of hotels was in the trend of decreasing. In addition, because of rough competition, hotels were below the break even point in 1998 and continued to worse till 2001. The travel agencies were in low profit with 2.08% in 2001. In a word, excessive supply over demand in tourism industry also responds to decrease of production factor quality and low efficiency of production factor utilization.

3. Conclusions

On the whole, we can conclude that China's tourism industry has the advantages in absorbing employment while it still remains on the phrase of intensive mode. China's tourism industry still remains on the phrase of high investment, high consumption and low efficiency. The research findings are the following:

Firstly, product quality of Chinese tourism industry got a little improved these years. Tourist satisfaction index increases these years while utilization index of product factors remains on the low level. Its room occupancy of star-level hotels decreases from 1994 to 1998 and a little increases from 1999 to 2002.

Secondly, production factor quality was in the trend of decrease from 1994 to 1999. However, it turned better from 2000 to 2002.

Thirdly, except for 2002, growth way of tourism industry still remains on the phrase of intensive mode. China's tourism industry depends more and more on capital investment.

Fourthly, operating quality is not well with frequency fluctuation and unreasonable industry structure. Excessive supply over demand leads to low efficiency of production factor.

In a word, contrast with its sharply expansive, China's tourism industry growth quality is worrisome. The results show that China's tourism industry should not only pay attention to its quantity expansion but also concern its growth quality. It should be transformed from intensive mode to extensive mode.

References

［1］Wang Jiye. Thoughts on Increasing Quality of Economic Growth［J］. *Macroeconomics*, 2000,(1)：11–17(in Chinese). 王积业 :《关于提高经济增长质量的宏观思考》[J]，《宏观经济研究》，2000 年第 1 期，第 11~17 页。

［2］Jefferson, Alan，Leonard Lickorish. *Marketing Tourism: A Practical Guide*［M］. Essex: Longman Group UK Limited 1988.

［3］Heath, Ernie, Geoffrey Wall. *Marketing Tourism Destinations: A Strategic Planning Approach*［M］. New York: John Wiley & Sons, Inc 1992.

［4］Song Zhenchun, Chen Fangying, Li Ruifen. A Cognition of Tourism Industry–A Discussion with Mr. ZHANG Tao［J］. *Tourism Tribune*, 2004,(2):76–79（in Chinese）. 宋振春、陈方英、李瑞芬:《对旅游业的再认识——兼与张涛先生商榷》[J]，《旅游学刊》2004 年第 2 期，第 76~79 页。

［5］Sara Nordin. *Tourism Clustering & Innocation–paths to Economic Growth & Development*

[M] . Sweden: European tourism research institute, 2003.15.

[6] Zhang Guangrui, Wei Xiaoan, Liu Deqian. *Green Book of China's Tourism: Analysis and forecast of Chinese tourism industry, 2003-2005* [M] . Beijing: Social Science Documentation Publishing House. 2005. 60,95 (in Chinese). 张广瑞、魏小安、刘德谦《2003~2005 年中国旅游发展：分析与预测》[M]，北京，社会科学文献出版社 , 2005，第 60、95 页。

[7] Liu Haiying. *Research on the Quality of Economic Growth in China* [D] . Doctor paper of Jilin University, 2005(in Chinese). 刘海英：《中国经济增长质量研究》[D]，《吉林大学》, 2005。

[8] Topic Group for Study of the Creativeness of Increasing Forms of Investment Performance of Travel Industry in Beijing. The Status Quo and Analysis of Investment and Operational Performance of the Three Key Industries of Travel Industry In Beijing [J] . *Tourism Tribune*, 2003, (2): 41-46 (in Chinese).《北京市旅游业投资效益增长方式的创新研究》课题组 . 北京市旅游业三大核心产业投资经营效益现状与剖析 [J]，《旅游学刊》，2003 年第 2 期，第 41~46 页。

On the Assessment of Economic Benefits of China's Listed Tourism Companies

Liu Liqiu, Zhao Liming, Duan Erli

(School of Management, Tianjin University, Tianjin 300072, China)

Abstract: Based on the study of current system of evaluating methods both at home and abroad, the paper chooses ten indexes including capacities for earning profits, solvency, operation and growth to construct management performance assessment index system of listed companies. Based on the annual data and using factor analysis, the paper makes an assessment analysis of China's 32 listed tourism companies in the hope to offer reference for listed companies to regulate assets, improve administrative structure and increase their value.

Key words: listed tourism company; EVA; factor analysis

Tourism, as a sunrise industry, has a broad market prospect. With social economic development, scientific-technological progress and transformation of people's values, tourism activities are changing and developing constantly. China has always been making tourism as the leader of the third industry, but China's existing tourism enterprises can not be among the forefront of the world no matter on scale or competitiveness. So that there should be a new, higher demands to further development of China's tourism enterprises. Therefore, a

[About authors] Ms. Liu Liqiu (1965-), Ph.D, associate professor of Management School of Tianjin University, the main research directions focus on the tourism management, technical and economic analysis, E-mail: lliqiu@126.com; Mr. Zhao Liming (1951-), professor of Management School of Tianjin University, doctoral tutor, the main research directions for regional economic sustainable development, enterprise management, market research; Ms. Duan Erli (1982-), master, Management School of Tianjin University, the research study focus on the project investment decision-making.

new definition of China's listed tourism companies and reevaluating them by their financial data will give a reference for helping listed tourism companies as a whole to enhance competitiveness and capability of resistance against risk.

1. Review of the cost–effective evaluation method of listed companies

Along with the continuous development of China's securities market, the number of listed companies has increased significantly. Different stakeholders have different choices and evaluation methods to operating performance of listed companies. How to assess the economic benefits of listed companies scientifically and rationally becomes one of the important topics both on China's securities regulatory institution and investors.

EVA (Economic Value Added), created by Stern & Stewart consulting company, refers to the difference between the capital gains and the cost of capital, specifically refers to the difference between business net profit after tax of enterprises and all investment capital (the sum of debt capital and private capital). EVA is a performance evaluation and incentive system, whose purpose is to enable company managers to maximize shareholders' value as their behavior criteria and to add economic value of companies' existing assets as a yardstick to measure the performance.

EVA evaluation method will make some necessary adjustments to the accounting information sources before calculation, in order to remove the impact of the evaluation from the accounting distortion, which will increase the complexities and difficulties in the calculation. Modern corporate governance structure theoretically requires owners to handover decision-making powers to managers. While on immature stage of management theory, technology, market and competitive environment in China, the coordination problems between owners and operators and the performance of the business involved in all aspects of the problem still cannot be resolved and reflected [1]. James Dodd and Shimin Chen, from the United States, did some empirical studies on the operating performance of 566 companies during 1983 to 1992, which could not prove that the EVA could provide more information than other indicators' did and this also means that it is difficult to prove that EVA is better than other indicators [2].

In 1990s, American scholars Robert Kaplan and David Norton created the Balanced Score Card for performance evaluation, which called BSC for short. The method which is stood by the side of corporate strategy planning and management is to use financial performance, customer, internal business process, learning and growth this four major aspects to comprehensive measure and evaluate the operating performance of companies. The Balanced Score Card has a fatal defect in that the non-financial indicators are difficult to quantify because BSC just provides a kind of method and builds up a theoretical framework, which is not an evaluation system of universal application. When application enterprises should be in accordance with their requirements and strategic management of the external environment to select the characteristics of different perspectives and targets for the design of the Balanced Score Card to decide indicators' weights and performance evaluation standards. There lack common standards of evaluation scale in the four aspects mentioned above. Those indicators not only need large quantities of information, but also need fully process in order to get their full value, which bring much higher demand for companies' information transmission system and feedback system. The Balanced Score Card only has relatively accuracy [3].

At present, "the Comprehensive Sorting of Chinese Listed Companies Management Performance", which is launched by China Securities Integrity Assessment Ltd. and China Securities News and has been publishing in China Securities News announced every year since 1996, gains a relatively high domestic reputation in China. Its evaluation method is as following: selecting Return on Net Worth, Total Assets Growth Rate, Total Profit Growth Rate, Debt Ratio, Current Ratio and All Capitalized Ratio as evaluation indexes. And each company's score for each indicators multiplied by the weight of them, and adding all of them, then the sum is the total score. Weighted score additive synthesis is one of the common methods in the comprehensive evaluation, but a prominent issue of this method is in it's subjective cognition in the process of weighting selection. While this method also can not considers degree of correlation of all the various indicators, that is, the duplication degree of every index which reflects its information. In fact there are some degree of relevance among various indicators.

Factor analysis method is a highly efficient statistical method which studies the dependence of the relevant internal matrix firstly and then does some structure treatment on the observation variables which are complicated in relationship, and finally ultimately reduces a few main factors (not related). The basic method is based on the size of the relationship to group the related variables, making the same group of variables more relevant and different groups of variables less relevant. Every group of variables represents a basic structure which is called the public factor or main factor.

There are p targets which are observable, $x_1, x_2, \cdots, x_p$. m factors which are not observable, F1, F2, ⋯ , Fm. The mathematical model of the factor analysis can be expressed as:

$$\left\{\begin{array}{l} x_1=a_{11}F_1+a_{12}F_2+\cdots+a_{1m}F_m+\varepsilon_1 \\ x_2=a_{21}F_1+a_{22}F_2+\cdots+a_{2m}F_m+\varepsilon_2 \\ \cdots\cdots\cdots\cdots\cdots\cdots\cdots\cdots\cdots\cdots\cdots\cdots \\ x_p=a_{p1}F_1+a_{p2}F_2+\cdots+a_{pm}F_m+\varepsilon_p \end{array}\right.$$

Note: $m<p$,

F_j (j=1, 2, …, m) is the public factors, and they are orthogonal each other.

ε_i (i =1, 2, ..., P) is the special factors, only playing roles to x_i.

a_{ij} is the load of public factor which is the load of the j factor on the i indicator, or a correlation coefficient. The load is lager, which means that the relationship of the i indicator and the j factor is more closer, otherwise is more alienated.

The goal of the factor analysis is to made the variable F_j controlling the original indicators and including more original informations, so as to get a clear observation and simplify observing systems and thus to overcome the defects such as the correlation of indicators and index's weighting made by people in the traditional evaluation.

Factor analysis gets the weight of indicators in the Composite Score by the data itself, which reflects the objectivity and fairness of the operating performance evaluation well. Although it will affect the final comprehensive evaluation function by choosing samples in different selection periods, samples' difference or samples of different industries, it will not reduce the comparability of the companies operating performance in the same sample or same period, which will also overcome the defect in traditional evaluation methods that when the weight is determined it will change a little. Therefore, using factor analysis to evaluate cost-effectiveness of listed companies has a wide application, both on a separate evaluation methods or a reference and a supplement of other evaluation methods [4,5].

2. The range of listed tourism companies in China

In the early establishment of China's securities market, relevant departments have not classified listed companies. Shanghai and Shenzhen Stock Exchange in accordance with their respective needs made a simple division. Shanghai Stock Exchange divided listed companies into industrial, commercial, public utilities and integrated utility as the four categories; Shenzhen Stock Exchange divided it into industrial, commercial, public utilities, financial and integrated utility as the five categories. There has been a great difference in the scope of listed traveling companies because of those simple categories for a long time. Particularly in recent years, the specific number of listed traveling companies has raised from some 20 to 30 in the same period in the different journals. The main reason for this situation, firstly, is that the tourism industry covers a wide range of scopes which includes food, housing, transportation, travel, procurement, entertainment. Secondly, the early primary business was tourism and with the development of the listed companies and the adjustment of industrial structure, the main operating structure changed. Thirdly, some listed companies were involved in tourism. Finally, the main projects of some companies have the features of tourism, such as the Zhongshi media in the Wuxi Ying Shicheng and the Oriental Pearl TV Tower in Shanghai. There appeared different definitions for the listed traveling companies because of the four reasons above.

According to the guidelines on listed company sector which was promulgated by the China Securities Regulatory Commission (CSRC) in April 2001, the Stock Exchange classified the listed company in terms of the consolidated accounts date that is audited by the accounting firm using the income proportion of listed companies as its classification criteria. The principles and methods of categories are, when the operating income ratio of a listed company's is greater than or equal to 50 percent, putted it into the corresponding categories of the business. When the company does not have the type of business operating income ratio greater than or equal to 50 percent, but the proportion of operating income in certain business is greater 30 percent than others, then put it into the corresponding categories of the industry. Otherwise, it is zoned as the comprehensive category. According to the guidelines, there are 32 listed tourism companies in Shanghai and Shenzhen exchanges in the paper, see Table 1.

Table 1   List of Listed Tourism Companies in China

StockCode	Company Name	StockCode	Company Name	StockCode	Company Name
600258	Shougang Shares	000033	XinDu Hotel	000835	SiChuanda
600138	Zhongqing Tourism	000069	HuaQiao City A	000610	XiAn Tourism
600358	Guolvlianhe	000524	OrientalHotel	000043	Shennanguang A
000428	Huatian Hotel	000721	S WestRestaurant	600749	TibetTourism
000430	S Zhangjiajie	000600	JianTou Energy	600593	DaLianShengDa
000888	Emeishan A	600754	JingJiangShares	000802	PeKingTourism
000978	GuilinTuorism	000691	ST Saidao	600515	S*ST YiTou
600654	Huang Shan Tuorism	600650	Jing Jiang Investment	000679	DaLian YouYi
000639	Kingde Develepment	000711	TianLun ZhiYe	000613	S*ST DongHai
000419	Chong cheng Share-holding	000501	E Wushang A	002033	LiJiang Tourism
600175	Meidu Share-holding	600873	Wuzhou Pearl		

The business scope of the listed traveling companies includes basically hotels, travel agencies, scenic spots and entertainment industries closely related to tourism, such as product development. The listed companies are the capital management companies, with the aim of maximizing the net assets field, and capital will be turned to the most profitable industries after the listed companies is constructed and raise funds. So it is difficult to define which industry the listed tourism company is included. The companies which are the first listing have the relative single business and also include other aspects of the tourism industry, even include in those companies outside the tourism industry after many years of operation. From the time of distribution of the listing, the listed companies were mainly hotels before 1995. In 1996 and 1997 scenic spots, travel agencies mainly. And a comprehensive configuration tourism elements companies mainly after 1998.

3. Economic efficiency evaluation of the listed tourism companies in China

3.1   The establishment of evaluation index system

Based on the above analysis, according to the study of the evaluation system at home and abroad, there are ten indicators of reflecting profitability, solvency, viability and growth capacity of the listed companies which can construct the perating performance evaluation index system of listed companies.

3.1.1   Profitability indicators

Profitability of the company's performance is essential. Therefore, we can describe it with net assets yield (X_1) and net profit of total assets (X_2). In addition to add the main business profit margins (X_3), because that there will be often instability for the total profit and net profit such as the trans-feration of the rights to use land, duty-free and so on. While the main business profit margins is to exclude these instability indicators, so as to reflect the company's true profitability perfectly.

3.1.2   Solvency indicators

Short-term liabilities: using cross-references of the current ratio (X_4) and liquidity ratio (X_5), and not just choose the current ratio commonly as a central target. Current ratio may reflect a short-term solvency, however, the higher current ratio does not mean that the companies have sufficient cash and deposits to debt repayment. How about the short-term solvency in the end should also use the liquidity ratio as reference, which is the ratio of liquidity ratio and current ratio after removing the inventory and prepaid expenses by the terms of the existing rule of the general principle of financial system or financial provisions.

Long-term liabilities: using equity-debt rate (X_6) rather than multiples of security interest. Although the two are the important indicators that reflect long-term liabilities' ability of the companies, the multiple security interest calculation needs to use an interest charges. China's profit report does not provide separate interest charges, but it is mixed in financial costs. External analysts can not accurately calculate the index, and the operable is poor. Therefore, the assets liability rate is selected.

3.1.3   Operating indicators

Because the two most important aspects are receivable accounts and inventory turnover efficiency, so we select the receivable accounts turnover rate (X_7) and inventory turnover rate (X_8).

3.1.4   Growth indicators

Considering the size and profitability, so we choose per share earnings growth (X_9) and total assets growth rate (X_{10}).

3.2   Empirical Analysis

In this paper, we choose the tourist section on listed companies as a sample and the number of samples (of 32) far exceeds the number (10) from the sample capacity, which meets the basic requirements of the factor analysis. In additional, because the companies come from same industry, which makes performance evaluation more scientific and reliable? All indicators are from the company's Annual Report in 2004. In addition that the rate of assets liabilities is the cost (that is, the smaller the better) indicator, other indicators are of efficiency in the established indicators system, so which should be tended to the same. In order to maintain the linear structure of data, the original target should be unsteadied by the contrary figures of assets liabilities rate, and other targets remain unchanged, so we will get the original target matrix after convergence.

3.2.1   Evaluation Standardization

In order to eliminate the non-comparability of targets resulting from different dimension units, using the formula:

$$x_{ij}^* = (x_{ij} - \bar{x}_j)/\sqrt{Var(x_j)} \text{ (i=1,2}\cdots\text{, n; j=1, 2, } \cdots\text{, p)} \tag{2}$$

Standardizing the pretreatmented data, and

$$\bar{x}_j = \left(\sum_{i=1}^{n} x_{ij}\right)/n, \sqrt{Var(x_j)}$$

$$= \sum_{i=1}^{n} (x_{ij} - \bar{x}_j)^2/(n-1), (j=1,2\cdots,p)$$

After this transformation, standared data is not in connection with the metric units of raw data. We will get the standardized matrix, that is $X=(X''_{ij})$

3.2.2　Related array of standardized matrix

According to the standardized matrix to calculate the correlation coefficient matrix $R=(r_{ij})_{p*p}$. From Table 2, the 10 indicators exist strong correlation each other, that is, there will be considerable overlap between information reflected by the various indicators and this analysis matches the original. According to the KMO value and the value of Bartlett, this also meets the test, so this can take advantage of the factors analysis in dealing with the multi-dimensional related indicators objectively.

Table 2　Related Array of Standardized Matrix

Compoent	RNW	CNPR	MBPR	CR	LR	DAR	STR	RT	EPSGR	TAGR
Return on Net Worth	1.000	0.836	0.096	0.111	0.126	-0.265	0.104	-0.674	0.583	0.080
Capital Net Profit Rate	0.836	1.000	0.197	0.357	0.373	-0.505	0.081	-0.536	0.527	0.327
Main Business Profit Rate	0.096	0.197	1.000	0.137	0.211	-0.005	-0.083	-0.045	0.077	0.182
Current Ratio	0.111	0.357	0.137	1.000	0.974	-0.483	-0.018	-0.157	0.120	0.497
Liquidity Ratio	0.126	0.373	0.211	0.974	1.000	-0.511	0.058	-0.135	0.146	0.463
Debt to Assets Ratio	-0.265	-0.505	-0.005	-0.483	-0.511	1.000	-0.232	0.237	-0.317	-0.135
Stock Turnover Rate	0.104	0.081	-0.083	-0.018	0.058	-0.232	1.000	-0.120	0.388	0.174
Receivables Turnover	-0.674	-0.536	-0.045	-0.157	-0.135	0.237	-0.120	1.000	-0.834	-0.200
Earnings Per Share Growth Rate	0.583	0.527	0.077	0.120	0.146	-0.317	0.388	-0.834	1.000	0.224
Total Assets GR	0.080	0.327	0.182	0.497	0.463	-0.135	0.174	-0.200	0.224	1.000

3.2.3　The related matrix eigenvalue and the cumulative contribution rate

In Table 3, the cumulative rate of the former four principal components is 80.412 percent, higher than conventional 80% of the value. Therefore, using the first four principal components to evaluate the 32 listed companies comprehensively. This will simplify the analysis structure. Also it is able to maintain sufficient accuracy. At the same time, using SPSS13.0 to output the rotated factor load matrix (see Table 4).

3.2.4　The economic explanation of factor analysis

From the corresponding the load factor matrix after being rotated, we can see that the coefficient of the net capital gains rate and the net profit rate of the total assets in the first principal component are positive and relatively large.

Table 3   Related Marix Eigenvalue

Component	Initial Eigenvalue			Extraction Sums			Rotation Sums		
	Total	% of Variance	Cumulative %	Total	% of Variance	Cumulative	Toatal	% of Variance	Cumulative %
1	3.828	38.279	38.279	3.828	38.279	38.279	3.008	30.078	30.078
2	2.046	20.464	58.744	2.046	20.464	58.744	2.605	26.052	56.130
3	1.177	11.766	70.510	1.177	11.766	70.510	1.250	12.501	68.631
4	.990	9.902	80.412	.990	9.902	80.412	1.178	11.781	80.412
5	.780	7.802	88.214						
6	.580	5.802	94.015						
7	.394	3.943	97.958						
8	.131	1.305	99.263						
9	.057	.573	99.836						
10	.016	.164	100.000						

Table 4 Rotated Load Factor Matrix

	Component			
	1	2	3	4
Return on Net Worth	0.917	0.060	-0.088	0.002
Capital Net Profit Rate	0.809	0.381	-0.080	0.096
Main Business Profit Rate	0.113	0.053	-0.161	0.819
Current Ratio	0.055	0.946	0.023	0.165
Liquidity Ratio	0.065	0.941	-0.114	0.189
Debt to Assets Ratio	-0.338	-0.674	0.904	0.331
Stock Turnover Rate	0.103	0.031	0.904	-0.109
Receivables Turnover	-0.855	-0.029	-0.174	-0.074
Earnings Per Share GrowthRate	0.794	0.037	0.445	0.068
Total Assets Growth Rate	0.069	0.466	0.388	0.552

Therefore, the first principal component portraits the profitability of the listed travelling companies, which we call it on the profit factor in this paper. The coefficient of the current ratio and liquidity ratio are both larger in the second principal component. So the solvency of listed tourism companies is mainly depicted by the second principle which is called on debt factor in this paper. The coefficient of the inventory turnover rate is larger and positive in the third principal component, by which the operating capacity of listed companies are depicted, known as the operating factor. The main business profit margins are positive in the fourth principal component, the remaining smaller or negative.

3.2.5   Principal Component Composite Scores

Using the Regression Analysis to obtain the single-factor score function, and using the contribution rate weight to construct the comprehensive evaluation function [6]:

$$\text{SCORE} = \sum_{i=1}^{n} a_i F_i \text{ (ai is the contribution rate)} \tag{3}$$

Bringing the standardized financial ratio data of the listed company into the linear expressions of the principal components, and then we can get the scores of every principal components and receive the comprehensive evaluation index according to the comprehensive evaluation index expressions(*). Obviously the higher the index, the enterprise's operational performance the better. On the contrary, the lower, the the enterprise's operational performance the poorer, according to their descending sort.

Table 5   Principal Component Evaluation Scores and Performance Rank

Company Name	Score	Rank	Company Name	Score	Rank	Company Name	Score	Rank
LijingTourism	4.602332	1	Dalianshengda	0.422025	12	Emei Mountian A	-0.14385	23
Jingjiang Investment	1.605575	2	S*ST EastSee	0.350082	13	Oriental Hotel	-0.14881	24
Tibet Tourism	1.384648	3	DaLianYouYi	0.309558	14	Tianlunzhiye	-0.29887	25
XiAn Tourism	1.317507	4	HuatianHotel	0.30089	15	S West Restaurant	-0.38141	26
Shenguangnan A	0.826047	5	GuiLinTourism	0.25486	16	XinDu Hotel	-0.4958	27
Kingde Development	0.651907	6	JiantouEnergy	0.235043	17	Tongcheng Konggu	-0.5251	28
Guolvlianhe	0.641956	7	WuZhou Pearl	0.198278	18	Peking Tourism	-0.99509	29
Jingjiang Shares	0.633566	8	Shoulv Shares	0.028253	19	ST Huandao	-1.26176	30
Huaqiao City A	0.537461	9	Huangshan Tourism	0.022684	20	Ewushang A	-3.66586	31
SiChuanshengda	0.449014	10	S Zhangjiajie	-0.02873	21	S*ST YiTou	-7.20923	32
Zhongqing Tourism	0.435161	11	Meidu Holdingshares	-0.05235	22			

From the distribution of scores, the highest is the Lijiang tourism and the lowest is the S * ST YiTou. Performance is basically a normal distribution. Some enterprises which are effective, financial structure reasonable and operation stable are clearly the top in the rank.

4. Conclusion

Through the analysis we can see that the economic efficiency of the listed tourism companies of our country has a larger heterogeneity. The ability is still premature using traveling capital of listed companies through capital markets operations and asset integration so as to increase the value of enterprises. Standardizing corporate governance of the listed tourism companies, improving financial situation and their internal control system through the reorganization of the high profits involved in growth industries and improving the scientific nature of the investment decision-making will achieve sustained and healthy development target of the listed tourism companies in the future.

References

[1] Bai Hua. Reflections on the EVA evaluation method [J] . *Technology and Management*, 2003, (2): 44–45(in chinese). 白华:《EVA 评价方法的几点思考》[J],《科技与管理》2003 年第 2 期，第 44~45 页。

[2] Chen S, Dedd J L. Economic value added(EVA TM): An empirical examination of a new corporate performance measure [J] . *Journal of Managerial Issues*, 1997, (9): 318–333.

[3] Dong Li.*The method of evaluating the operation performance factors of the Private–listed companies and the analysis of empirical research* [D] . Wuhan: Wuhan University of Technology, 2005:6–12(in chinese). 董理:《民营上市公司经营业绩因子分析评价法及实证研究》[D],武汉，武汉理工大学 , 2005，第 6~12 页。

[4] He Xiaoqun. *Multivariate statistical analysis* [M] . Beijing: Chinese People's University Press, 2004. 168–172(in chinese). 何晓群:《多元统计分析》[M],北京，中国人民大学出版社 , 2004，第 168~172 页。

[5] Han Bing. Comprehensive evaluation of the Corporation research based on the factor analysis [J] . *Business research*, 2004, (13) : 55–56(in chinese). 韩冰:《基于因子分析的证券公司综合评价》[J],《商业研究》2004 年第 13 期，第 55~56 页。

[6] GU Wenjiong. Financial evaluation of listed agricultural companies by the factors analysis [J] . *Anhui University Journal*, 2005, (5): 136–139(in chinese). 顾文炯:《用因子分析法对农业上市公司进行财务评价》[J],《安徽大学学报》2005 年第 5 期，第 136~139 页。

Evaluation on Regional Tourism Competitiveness: Index System and Methodology

Zhang Meng

(School of Business Administration, South Western University of Finance and Economics, Chengdu 610074, China)

Abstract: To reasonably construct index system and scientifically evaluate regional tourism is the precondition of regional tourism competitiveness comparison, situation analysis and making countermeasure proposals. It's also the important content of regional tourism competitiveness theory. The paper takes documentary review as the study base combined with regional tourism development practice and competitiveness development character to build a regional tourism competitiveness synthetical index system. Furthermore, the paper lays discussion and choices of evaluation research methods.

Key words: tourism competitiveness; index system; evaluation research methods

1. Review and comments on previous research

It's not difficult to find that there are mainly two kinds of evaluation approaches on competitiveness when we take a review on previous research.

[About author] Ms.Zhang Meng, professor of School of Business Administration of Southwestern University of Finance and Economics, Ph. D. of economics, with research interests centred on issues related to the study of regional tourism competitiveness and cross-cultural tourism.

1.1   Single-factor based evaluation and analysis

Single-factor based evaluation and analysis is the method that takes one factor as evaluation index to reflect competition situation of research objects. Representative methods and indexes are as follow: profitability, market share, output capacity, export & import data, etc.

When competitiveness evaluated by single-factor based methods, on account of less related data, it's simple and easy to handle. But for competitiveness is a capacity system, the single-factor evaluation method is lack of comprehensiveness.

1.2   Multi-factor based evaluation and analysis

For competitiveness is evaluated by combined factors, the method also is called comprehensive approach. For example, World Competitiveness Year Book (WCY) released by the International Institute for Management Development (IMD), mainly uses comprehensive approach to evaluate competitiveness. Michael E. Porter's evaluation of industry competitiveness and enterprises competitiveness also is a comprehensive evaluation method which takes macro environmental factors, industry & policy conditions and government & enterprise factors into account. Many China's domestic scholars who also take the method, such as Zheng Gang & Jiang Chunlin use for reference of domestic & overseas advanced theories and measures to build a systematic evaluation index that includes obvious competitive capacity, potential competitive capacity, competition environment and competition situation [1].

On the other hand, some guilds and consulting companies take surveyor's pole approach to evaluate and judge competitiveness, actually, the method is a multi-factor evaluation approach. Generally, the method firstly locates key factors to competitiveness and then, probes the factors' best usage for compared nations or enterprises. At last, taking the object compared with the world wide best practice and thus shows the competitiveness gap.

The advantage of multi-factor based analysis is that it can be as comprehensive as possible, however, the method needs get factors combined to educe competitive situation. When the approach is carried out, weighing to the factors will affect the final results.

With review on tourism competitiveness literature in recent years, multi-factor evaluation still is the main analysis approach. China's Tourism International Competitiveness Systemic Evaluation (by ZhuYinggui, WanXucai), takes tourism resources & product, social economic status, international tourism output and other related factors into account [2] ; Theory and Method on Urban Tourism Competitiveness Research (by Caoning, Guoshu), the index factors include core attractions, fundamental factors, supportive factors, qualificatory factors and managerial innovation factors [3] ; GuoLufang's index system includes tourism competitive strength, potential tourism competitive strength and future tourism development prospect to value international tourism competitiveness [4] ; Li Guozhu's regional tourism appraising model includes dimensions competitiveness, market competitiveness, enterprises competitiveness, primary competitiveness, development competitiveness, environmental supportive competitiveness [5] , etc.

From the review of previous disquisitions of tourism competitiveness evaluation, there still is insufficiency as follows:

First, index system is lack of integrality. mainly for lacking of in-depth study on related factors and results in insufficient in constructing index system.

Second, indexes' choosing is lack of actual practical value, especially, when it comes to samples. For indexes substitution, index system construction is lacking of general practical value.

Third, research methods are on the primary stage, many of them are repetitious studies. For example, when taking multi-factor combined, The Analytic Hierarchy Process (AHP) is mainly used and that results in the evaluation and analyses are somewhat subjective.

Finally, all the previous researches took static analysis with cross-section data. As a matter of fact, the factor s' importance will be changing when time goes by and some of them will be less important. As for this, it's seldom for scholars to make use of timing panel-data in analysis and evaluation to forecast developing trends.

2. Regional tourism competitiveness evaluation: indexes choosing and index system constructing

2.1 Matters should be taken into account in tourism competitiveness evaluation

2.1.1 Regional tourism competitiveness is the capacity that makes the region provide the market with tourism product and service constantly.

It has the feature of profitability and durative. The terminating indication of competitiveness is the market share and profitability of the tourism product; furthermore, it stands for actual competitiveness comparison results. Thus, competition results should be the beginning of regional tourism competitiveness evaluation. Competitiveness is not only a comparative concept but a dynamic one; as a result, competition object's constant competitiveness is related to decisive or effective factors. All in all, we need to evaluate the decisive and effective factors besides the objective description of industrial competition results when it comes to regional tourism competitiveness evaluation. Followed by this disquisition, the construction of regional tourism competitiveness evaluation index system mainly includes two kinds of indexes: one reflects results of competitiveness, i.e. outer indexes; the other explains the reasons of having competitiveness, i.e. inner indexes.

2.1.2 The aim of the dissertation is to construct tourism competitiveness empirical analysis evaluation model, so the ideal indexes must be simplified and corrected during the construction of the index system, which should be scientific and practical, i.e. directed by competitiveness theoretical analysis (scientific) and at the same time, based on practice (convenient for analysis, evaluation and observation).

2.1.3 For actual decisive factors of industrial competitiveness are those vital factors that result in differences among competition objects, the dissertation's rudder is taking emphasis on those vital decisive factors and laying stress on key points.

2.1.4 Comparison among regions, to ensure the coherence of the same index's connotation, extension, arithmetic and timetable, etc.

When puts to idiographic evaluation, standardization, unitary character and maximum or minimum of indexes data must be carried out so as to makes the data comparable under no dimension condition.

2.1.5 Mainly taking hard indexes.

Hard Indexes generally refer to measurable indexes, which can be gotten from statistic data. On the contrary, Soft Indexes are hard to measure and mainly are the understanding and judgment indexes of industrial competitiveness, the soft indexes can be gotten from questionnaires. On account of reliability and acquirability, the paper mainly chooses the hard indexes.

2.2 Structure and content of the index system

Directed by the above principles, the thesis builds a regional tourism competitiveness evaluation index system taking key effective factors into account. Presented as table 1, the index system includes 4 kinds of 55 indexes. Hereinto, F_i stands for the four groups of essential competitiveness evaluation indexes, X_{ij} stands for direct observation indexes.

2.2.1 Indexes of regional tourism market competitiveness(F_1)

Market competitiveness indexes are the most direct reflection of tourism

Table 1   Regional tourism competitiveness evaluation index system

Tourism synthetical competitiveness	Market competitiveness (F1)	x_{11}—ratio of profit to output value x_{12}—market share x_{13}—number of international tourists by region(10000 person-times) x_{14}—tourism takings(10000yuan) x_{15}—foreign exchange earnings by region(USD million) x_{16}—tourism realize pre-tax profits per labor(10000yuna/person) x_{17}—original valued of fixed assets per labor(10000yuan) x_{18}—asset-income ratio
	Core competitiveness (F2)	x_{21}—resources conditions x_{22}—resources abundance degree x_{23}—tourism overall productivity(10000yuan/person) x_{24}—star-hotels' room rent rate(%) x_{25}—overall technical efficiency x_{26}—fixed asset turnover x_{27}—training rate of tourism employees x_{28}—degree of concentration of travel agencies x_{29}—degree of concentration of hotels
	Essential competitiveness (F3)	x_{31}—ration of high-grade high way(%) x_{32}—density of highways and railroads x_{33}—volume of traffic x_{34}—internet users per 10000 persons x_{35}—mobile phone users per 10000persons x_{36}—telephone service traffic (a hundred yuan/10000persons) x_{37}—l.d.tel. usage volume(ten thousand/10000person) x_{38}—number of star-hotels x_{39}—number of urban service facilities x_{310}—number of museums x_{311}—amount of water supply by region(ten thousand stere) x_{312}—amount of electric power consumption x_{313}—environment protection employees(person) x_{314}—urban sewage disposal capacity per day(ten thousand stere) x_{315}—public green-land area per person(centiare) x_{316}—qualification rate of industrial sewage discharge(%) x_{317}—disposition rate of industrial solid wastes(%) x_{318}—number of medical institution beds per 10000persons

Continued Table 1

Tourism synthetical competitiveness	Essential competitiveness (F3)	X_{319}—ratio of illiteracy & semiliterate among 15 year-old and older people(%)
		X_{320}—number of people received college and upper education
		X_{321}—number of college students
		X_{322}—ratio of local fiscal expenditure of three items to expenditure on science and technology(%)
		X_{323}—per capita GDP(yuan)
		X_{324}—amount of household consumption by region(yuan)
		X_{325}—average yearly amount of townsman disposable income(yuan)
	Systemic competitiveness (F4)	X_{41}—contribution rate of import
		X_{42}—contribution rate of export
		X_{43}—ratio of foreign investment
		X_{44}—ratio of GDP to local fiscal income
		X_{45}—degree of infrastructure construction investment
		X_{46}—ratio of fiscal expenditure of urban maintenance(%)
		X_{47}—ratio of state-owned star-hotels(%)
		X_{48}—ratio of Joint-stock star-hotels(%)
		X_{49}—ratio of foreign investment star-hotels(%)
		X_{410}—net sales amount of books by region(10000yuan)
		X_{411}—ratio of fiscal expenditure of culture & sports and broadcast
		X_{412}—number of public libraries
		X_{413}—amount of art show-place by region

competitiveness level, mainly including production market competitiveness, consumer-origin market competitiveness and industrial efficiency.

Production market competitiveness indexes are mainly reflected by profitability and market share. Profitability index reflects the capability of gaining profit and fortune, is the direct reflection of regional tourism competitiveness. Profitability indexes include: X_{11}-production profit margin; X_{12}(tourism market share) =regional tourism income/overall national tourism income, the index exposes regional tourism's capacity within regional competition, is the important index reflecting industry expanding ability. Generally, relatively higher market share means a better situation in industrial competition.

Tourism consumer-origin market is the precondition of the existing and development of tourism. Thus, regional tourism market competition mainly is how to attract and expand tourist-origin markets; furthermore, it has become a key point of tourism market competition. Tourist-origin market competitiveness indexes are reflected by X_{13}, X_{14}, and X_{15}.

Industrial efficiency somewhat reflects tourism competition quality. Industrial efficiency indexes include X_{16}, X_{17} and X_{18}.

2.2.2　Indexes of regional tourism core competitiveness evaluation (F_2)
The follow will focus on competitiveness of tourism attractions, enterprises, industrial structure to explore indexes F_2.

Competitiveness of tourism attractions is vital for improving regional tourism competitiveness through forming regional characteristic products, gaining relatively higher differential advantage. The dissertation takes X_{21} and X_{22} to evaluate. Thereinto, resources conditions(X_{21}) refer to the quality and value of tourism resources, which include the value of aesthetics, history, science and the degree of scarcity. For this, the paper chooses to be based on world heritage (cultural, natural and the both), state level scenic spots and province level scenic spots that occupy dominant position among Chinese tourism resources to calculate X_{21} by region. Resources abundance degree (X_{22}) means the intensity of resources within the region. Generally, the resources, arranged into lines or networks are more valuable than those can not be.

Competitiveness of enterprises is the capacity that can more effectively provide the market constantly with products than other competitors on competition markets; furthermore, it can gain the enterprise the capability of sustainable development. The capacity is mainly decided by the competitive advantage forming through the optimization of the enterprise owned essentials and the favorable interaction between the enterprise and its outer environmental factors. Thus, the thesis uses operating & management level and human resources condition, etc, to stand for enterprises competitiveness. Among them:

X_{23} stands for enterprises management level, it presents the production efficiency. X_{23}=tourism value added/number of tourism employees of year-end level. Besides that, X_{24} also is an index standing for management level.

Moreover, overall technical efficiency (X_{25}) can be used to evaluate whether enterprises' production is effective. "If under the circumstance that does not decrease others' output (or increase others' input), technically, can not increase any output (or decrease any input), the feasible input-output vector be called technical effective"[i]. The bigger of the value of X_{25} is, the better the level of resources allocation is, and the higher the output is. Thus, the index reflects the promotion of production efficiency because of innovation of business management, technology and system.

Asset turnover stands for enterprises operating level. Tourism enterprises asset turnover=net main business taking average amount of asset, however, China Tourism Statistical Yearbook only presents the data of fixed assets, the paper takes fixed asset turnover (X_{26}) as the substitute. On the other hand, main business profitability also is an index standing for tourism enterprises financial condition. The paper has taken ratio of profit to output value as the profitable ability index, on account of avoidance of repetition; do not list it to stand for operating level.

Human resources condition is the key factor to the promotion of enterprises competitiveness. Enterprises potential competitiveness is mainly rested with the abundance degree of human resources. Human resources are the base of innovation of enterprise value, products, management and technology. The evaluation index on human resources mainly takes the ratio of trained employees of all tourism employees (X_{27}). The index refers to there are how many employees who received whatever training projects during the past year.

Industrial structure competitiveness is mainly expressed by the degree of industry concentration. According to industrial organization theory, industry concentration is the key evaluation index to the competition situation of one industry or market. Hereon, the thesis chooses to take hotels and travel agencies as indexes reference objects, i.e. x_{28}—degree of concentration of travel agencies, x_{29}—degree of concentration of hotels. The lower the industrial concentration degree is, the weaker the dominant power of big business is, and the higher the competition degree is. It more likely results in cut-throat competition and resources waste. If one region has tourism competitiveness, the regional tourism should have comparatively higher industrial concentration, i.e. among the enterprises; there are certain numbers of big business. If there are no such enterprises as support, it's impossible for the regional tourism to gain industrial competitiveness nation-wide, not to say world-wide. Of course, the concentration is not the higher the better, however, confronted with the "smallness" and "scatter" situation of current Chinese tourism market, the concentration should be higher.

Some scholars also take the number of tourism enterprises as the index to evaluate tourism scale competitive advantage. They claim that one industry which has competitiveness in the region should be the industry of certain amount and scale; furthermore, that character will be helpful for the saving of operating cost and the improvement of management efficiency. But in fact, scale competitiveness of tourism is mainly exposed by the scale of single tourism enterprise up to now. For that, the paper does not take enterprises numbers index into account.

2.2.3 Indexes of regional tourism essential competitiveness evaluation (F₃)

Essential competitiveness has supportive effect on the improvement of regional tourism competitiveness, which is the base of regional tourism sustainable development. The indexes include the competitiveness of infrastructure, environment, human resources and synthetical economic power.

Infrastructure competitiveness chooses traffic condition, communication, scale of entertainment establishments, energy resource, etc. as observation points. Thereinto, traffic condition is one of the basal factors that affect regional tourism development. It not only has influence on the attractiveness for tourists but how tourists arrive at. The disquisition takes x_{31}— ration of high-grade high way, x_{32}— density of highways and railroads and x_{33}— volume of traffic to stand for the indexes; communicational indexes include x_{34}— internet users per 10000 persons, x_{35}— mobile phone users per 10000persons, x_{36}— telephone service traffic and x_{37}—l.d.tel. (Long distance telephone) Usage volume (ten thousand/ 10000person); the scale of entertainment establishments refers to the overall reception capacity, which includes the establishments of singing & dancing, P.E & setting-up, gaming, knowledge and others. The paper takes indexes of x_{38}— number of star-hotels, x_{39}— number of urban service facilities and x_{310}— number of museums to substitute; energy resources indexes are x_{311}— amount of water supply by region (ten thousand stere), x_{312}— amount of electric power consumption.

X_{313}, X_{314}, X_{315}, X_{316} and X_{317} are taken to evaluate entironment competitiveness.

The paper chooses the country diathesis to express human resources competitiveness. The indexes are the five as follow: X_{318}— number of medical institution beds per 10000persons, X_{319}— ratio of illiteracy & semiliterate among 15 year-old and people older (%), X_{320}— number of people received college and upper education, X_{321}— number of college students, X_{322}— ratio of local fiscal expenditure of three items to expenditure on science and technology (%).

Synthetical economy competitiveness can be disclosed by X_{323}— per capita GDP (yuan), X_{324}— amount of household consumption by region (yuan), X_{325}— average yearly amount of townsman disposable income (yuan). Per capita GDP means the ratio of one region's gross domestic product (GDP) with its population within certain period, which is the important index to reflect social production and productivity. Amount of household consumption by region is the reflection of the scale and level of consumption, which reflects one region's economy development level during certain period from consumption angle. X_{325} means the real income of household after the deduction of income tax, which discloses the potential consumption ability of regional townsman, at the same time, also can be the reflection of the attractiveness and domination over outsides.

2.2.4   Indexes of regional tourism systemic competitiveness evaluation (F_4)

F_4 is the combined index that reflects one region's governmental management level and regional innovation capacity, which mainly are legal system environment, government management level, systemic innovation ability and regional cultural influence ability. On account of the difficulty of getting data, however, the paper takes the follow as substitute.

2.2.4.1   **Taking economic attractiveness indexes as the substitute of legal system environmental competitiveness indexes.**

The economic attractiveness indexes include x_{41}—contribution rate of import, x_{42}—contribution rate of export, x_{43}— ratio of foreign investment. The higher attractiveness of one region means the better of the region's legal systematic completion and operating order. At the same time, it also indicates that administrative approval system of local government is clear, administration ability is strong; furthermore, there is a fair competition environment for enterprises. Hereinto, X_{43}=the ratio of the actual foreign investment amount by region to the overall capital forming amount by region.

2.2.4.2   **Taking x_{44}—ratio of GDP to local fiscal income, x_{45}—degree of infrastructure construction investment, ratio of fiscal expenditure of urban maintenance (%) to substitute the competitiveness of government managerial ability.**

Among them, x_{44} means the ratio of fiscal income within budget that local government realizes to current period's local GDP, which reflects the concentration degree of newly increased fortune to the government. The amount of fiscal income, however, is vital to the governmental influence and domination on local economy. x_{45} and x_{46} indicate how strong the emphasis taken by local government on tourism development. Although infrastructure construction and urban maintenance are not wholly for the need of tourism development, the precondition of local tourism development, especially for tourism distributing-center, is the infrastructure construction condition, such as traffic condition and the construction of the city itself.

2.2.4.3   **Taking x_{47}—ratio of state-owned star-hotels (%), x_{48}—ratio of Joint-stock star-hotels (%) and x_{49}—ratio of foreign investment star-hotels (%) to stand for systematic innovation capacity.**

Hereinto, x_{47} is to evaluate the resources domination of government through property right. If government operates what should be market-oriented, the result might be the damage of go-aheadnism and investment impulse; furthermore, state-owned enterprises system is not the system chosen by the market, which commonly is inefficient. The paper takes the ratio of state-owned star-hotels to evaluate the market-oriented degree of one region. If the ratio is lower, it means the decrease of the direct using and control over economic resources by local government; and also, the degree of market-oriented resources allocation, i.e. the degree of market-oriented is up-rising. The indexes x_{48} and x_{49} reflect the developing speed of market-oriented, non-state-owned economic sections. The bigger the indexes data are; the smaller the direct using and control degree are. They also stand for the scope that market taking influence on is expanding. All in all, the rapid development of non-state-owned segments is the important symbol that market-oriented degree is up-ring.

2.2.4.4   Taking x_{410}—net sales amount of books by region (10000yuan), x_{411}— ratio of fiscal expenditure of culture & sports and broadcast, x_{412}—number of public libraries, x_{413}—amount of art show-place by region to stand for cultural influential competitiveness

Cultural influential competitiveness mainly refers to the coherence, orientation, encouragement and impulse of all kinds of cultural factors that are born with the procedure with the social and human comprehensive development. The above four indexes can disclose the cultural open degree of one region by broad extent, accordingly, disclose indirectly the regional cultural carve-out spirit, innovation atmosphere and reciprocation personal integrity.

3. Methods choice of synthetical evaluation of regional tourism competitiveness

3.1   The static evaluation of regional tourism competitiveness–based on cross–section analysis

Tourism competitiveness evaluation system is a multi-indexes comprehensive evaluation system, which usually takes weighted mean method assigning value to indexes; thus, assigning value is vital. At present, academia mainly uses the method of Analytic Hierarchy Process (AHP). AHP is a method of subjective weighing, the importance of indexes is judged according to experts' opinions. For the differences of knowledge, experiences and personal value standards among the experts, the judgments on the importance of indexes will be out of coherence and this would likely results in subjective value assigning. Because of the pertinences among some indexes, besides, it will bring inter-superposition &

interferences. The result is the deviation of weighing to the indexes. In order to avoid the defects of subjective value assigning and get not only objective & reasonable but exclusive result of comprehensive evaluation, the paper tries to take advantage of factor-analysis approach to evaluate and analyze the cross-section of the indexes.

Factor-analysis approach is the statistic method combining several variables x_1, x_2..., x_p (the random variables that can be observed, i.e. apparent variables) to a few factors F_1, F_2, ..., F_m (potential variables that can not be observed); furthermore, the method could disclose the pertinences between the index and its effective factors. The basic idea is to group variables according to their pertinences and make it higher pertinences of the variables of the same group. Every group of variables stands for an essential structure, which is called public factor in factor-analysis approach. The structure can reflect an aspect or dimensionality of the problem. Through taking the variance contribution ratio of several public factors as weighing to construct synthetical evaluation function, it can simplify multitudinous primary variables and effectively deal with superposition among the indexes. The evaluation is fairly objective.

Because there are 55 fundamental variables in the paper's indexes system, if carrying out factor-analysis on so many variables might arise the confusion of factors' meaning and the difficulty of explaining factors; thus, the practical value of factor-analysis approach can not be realized. For the above, in order to more clearly present the relationship between the factor and its variables, the thesis only puts forward factor-analysis on the fundamental variables of "market competitiveness" and "core competitiveness" indexes, as for the fundamental variables of "essential competitiveness" and "systematic competitiveness" indexes, only puts on main-part-analysis.

Just as factor-analysis approach, main-part-analysis approach aims at reducing dimensionality. Their applied targets, however, are not the same. Main-part-analysis approach usually is taken to bring comprehensive analysis & evaluation, taxis and variable filtration. As for factor-analysis approach, it can be used to make comprehensive evaluation, but what more is to make classified evaluation, comparison and analysis on samples and variables. The aim of the paper choosing to bring factor-analysis approach on market competitiveness and core competitiveness, is to put in-depth research on the structure and character of effective factors of the two kinds of indexes; furthermore, as to the indexes of essential competitiveness and systematic competitiveness, it only needs to bring comprehensive evaluation and taxis, main-part-analysis approach is a better choice.

The calculation procedure of the above two approaches is relatively complex. It will be quite convenient, however, by using the FACTOR module of SPSS (statistic software).

3.2   Regional tourism competitiveness dynamic evaluations–the analysis based on panel data

In order to more clearly observe the changing character and developing trend in timing sequence, the paper tries the method of taking panel data to put comprehensive analysis and evaluation on the sample 55 indexes.

So called Panel Data refer to the sample observation data getting from the sections selected on timing sequence, i.e. panel data can be the two dimensional reflection of character and law of variables both temporal and sectional.

There are many panel data analyses methods. The dissertation goes like the following. First of all, choosing continuous yearly data (more than five years is better) of certain sample (55 indexes) to analyze and evaluate. And then, combining the evaluation results together to compare and evaluate. At last, disclosing the dynamic developing character and future development trends of regional tourism competitiveness based on the first and second steps.

3.3   Clustering analysis on regional competitiveness

The function of clustering analysis is to build a grouping method, which is an approach grouping a set of samples or variables according to their or close or far-off essential relationsii. As long as the sample or the variables have numeric character, clustering analysis approach will take effect. The approach takes every sample as one group and then is to reduce number of groups by consolidation according to intervals among samples and its definition and finally to realize the aim of clustering analysis.

The paper takes market competitiveness, core competitiveness, essential competitiveness, systematic competitiveness as clustering variables bringing clustering-analysis approach to analyze tourism competitiveness by region. Hoping to recognize the actual conditions of tourism comprehensive competitiveness by region and furthermore, to find out the reasons how the differences of regional tourism come into being.

References

[1] Zheng Gang, Jiang Chunlin. Assessment index system research on the international

competitiveness of regional industry〔J〕. *Scientific Management Research*, 2001, (6): 24~27 (in Chinese). 郑刚、姜春林:《区域产业国际竞争力评价指标体系研究》〔J〕,《科学管理研究》2001 年第 6 期, 第 24~27 页。

〔2〕Zhu Yinggao, Wan Xucai. Comprehensive assessment of international competitiveness in Chinese tourism industry〔J〕. *Human Geography*, 2005, (1): 57~61 (in Chinese). 朱应皋、万绪才:《中国旅游业国际竞争力综合评价》〔J〕,《人文地理》2005 年第 1 期, 第 57~61 页。

〔3〕Cao Ning, Guo Shu. Research theory and methodology on competitive power of metropolitan tourism〔J〕. *Social Scientist*, 2004, (3): 85~88 (in Chinese). 曹宁、郭舒:《城市旅游竞争力研究的理论与方法》〔J〕,《社会科学家》2004 年第 3 期, 第 85~88 页。

〔4〕Guo Lufang. Thoughts on international competitive power of Chinese tourism〔J〕. *Tourism Science*, 2000, (2): 12~15 (in Chinese). 郭鲁芳:《关于我国旅游业国际竞争力的思考》〔J〕,《旅游科学》2000 年第 2 期, 第 12~15 页。

〔5〕Zhu Shunlin. Comparative analysis of technology efficiency of regional tourism industry〔J〕.*Reform of Economic System*, 2005, (2):116~119 (in Chinese). 朱顺林:《区域旅游产业的技术效率比较分析》〔J〕,《经济体制改革》2005 年第 2 期, 第 116~119 页。

〔6〕Li Guozhu, Shi Peiji, Guo Xiaodong. A study of competition-oriented strategy planning of regional tourism development:a case study of Gansu Province〔J〕. *Tourism Science*, 2005, (3):26~33(in Chinese). 李国柱、石培基、郭小东:《基于竞争力导向的区域旅游发展战略规划研究——以甘肃省为例》〔J〕,《旅游科学》2005 年第 3 期, 第 26~33 页。

〔7〕Luo Jiyu, Xing Ying. *Predication and analysis methods in economical statistics*〔M〕. Beijing: Tsinghua University Press, 1990.129 (in Chinese). 罗积玉、邢瑛:《经济统计分析方法与预测》〔M〕, 北京, 清华大学出版, 1990, 第 129 页。

Development of China's Forest Parks and Forest Tourism*

Li Shidong[1,3], Chen Xinfeng[2], Deng Jinyang[4]

(1. The General Office of the State Forestry Administration, Beijing 100714, China; 2. The Forest Park Office of the State Forestry Administration, Beijing 100714, China; 3. IGSNRR, Chinese Academy of Sciences, Beijing 100101, China; 4. Recreation, Parks, and Tourism Resources Program, West Virginia University, USA 26506)

Abstract: This paper has systematically examined the development trajectory of China's forest parks and forest tourism industry based on a comprehensive analysis of the change of four indicators–number, size, visitation, and tourism revenue relating to forest parks as well as an analysis of the change of the same four indicators in relation to different provinces (autonomous regions and municipalities), planning regions, and four great regions. The results indicate that China's forest parks have experienced four development stages, including the beginning stage, the rapid development stage, the steady development stage, and the rapid maturity stage. These four stages were featured by the accelerated increase in park number and area temporally, the park distribution shift spatially, the match between park visits and tourism revenues, and the increasing importance of forest tourism in the whole tourism industry of

[About authors] Mr. Li Shidong (1966-), senior engineer at the professor level, Ph.D., research areas on ecotourism and ecological assessment. E-mail: eastworld@sohu.com; Mr. Cheng Xinfeng, (1964-), senior engineer, Ph.D., research areas on ecotourism and forest tourism. E-mail: cxf8915@sina.com.; Mr. Deng Jinyang (1966-), professor, Ph.D., research areas on ecotourism and forest tourism. E-mail: djinya@yahoo.com.

* An earlier version of this paper was presented at the 2006 China-USA Tourism Forum held in October 2006 in Changsha. Data on China's forest parks and forest tourism development are from the Forest Park Office of the State Forestry Administration.

the country. It predictes that China's forest parks will still remain in the rapid maturity stage during the Eleventh Five-Year Plan Period. Also, it is estimated that: 1) by 2010, the total number of forest parks in the country will jump to 2800 from the current 1928; 2) the proportion of forest parks in western China will increase significantly with a gradual growth in the proportion of forest tourism revenues; 3) the number of park visitors and corresponding direct tourism revenues will grow at an average annual growth rate of 20% or so, with park visitation at 400 million visits and forest tourism revenues at 20 billion yuan by 2010; and 4) the forest tourism industry will increasingly play an important role in China's tourism industry, and by 2010, annual park visits and total forest tourism GDP will account for approximately 19% of total national tourist visits and about 13% of the total national tourism GDP.

Key words: forest park; forest tourism; development trajectory; time and space distribution; tourist visits; tourism revenues.

China is among countries that have the most abundant forest scenery resources. Stretching from the shore of the East China Sea to the Tibetan Plateau, also known as the "roof of the world", and from tropical rain forests to cold coniferous forests, China has forested lands of 280 million hm^2 that feature colorful forest landscapes. These landscapes are integrated with other spectacular natural scenery such as unique karst landforms, majestic icy glaciers, wonderful lakes and islands, and stunning valleys and waterfalls, as well as deep cultural deposits, which are made up of the most important components of China's natural and cultural heritage. In order to scientifically protect and actively utilize its abundant forest scenery resources, China began to develop forest parks in the late 1970s and early 1980s, and established its first national forest park-Zhangjiajie National Forest Park in 1982. At present, China has developed a forest park system that spreads out across 31 provinces, autonomous regions, and municipalities (excluding Hong Kong, Macao, and Taiwan, hereafter the same) that consists of national forest parks, provincial forest parks, and local forest parks, with national forest parks being the backbone of the whole system. The development of China's forest park system has played an essential role in the protection of the country's natural and cultural heritage, and has also promoted the development of China's forest tourism industry, which, to a large degree, has met the increasing public demand for outdoor recreation. In order to provide a scientific base for the future construction of forest parks and the development of the forest tourism industry, we collected a large amount of data on the aforementioned four indicators and systematically examined the changing patterns of these indicators.

1. Development stages of China's forest parks

China's forest park system is featured by periodic development as clearly evidenced by the increasing patterns of national forest parks in both number and size[1]:

1.1　The beginning stage (1982–1990)

Since the establishment of the Zhangjiajie National Forest Park in 1982, China has established a number of other national forest parks since then, including Tiantong and Thousand Island Lake National Forest Parks in Zhejiang province, Louguantai National Forest Park in Shaanxi province, and Liuxihe National Forest Park in Guangdong province. This time period is characterized by the experimental development of forest parks with little perceived influence, with recreational functions of forest parks not being widely recognized by the public. Consequently, the forest parks developed slowly with the creation of only 16 national forest parks, or on average, less than two national forest parks opening annually. However, these parks were constructed jointly by the Ministry of Forestry (renamed as the State Forestry Administration in 1997) and relevant provinces with more than 20 million yuan invested from the central government. Today, most of these early established national forest parks have become famous tourism destinations both domestically and internationally.

1.2　The rapid development stage (1991–1993)

This time period is very unique for four reasons. First, the central government made a decision to develop the tertiary industry in full swing as a response to Deng's Xiaoping famous speeches in the spring of 1992 in southern China. As a result, the role of tourism was recognized by the public and society. Second, with the increasing deterioration of the so called "two crises" associated with forestry (i.e., resource crisis and economic crisis), the long standing practice of timber harvesting as the single industry in most forest farms needed to be changed. Third, after a time of constructing forest parks and developing forest tourism, the resulting economic, ecological, and social benefits, as well as the role that forest tourism played in promoting local and regional economic development, have been acknowledged by all walks of life. Finally, the Ministry of Forestry held a national roundtable conference on forest parks and forest tourism in 1992 in Dalian, which required national forest farms to set aside those places with a higher amenity of forest environment, rich biological resources, and a higher concentration of natural and cultural attractions as forest parks. This call stimulated a new wave of forest park construction across the country, resulting in a total of 218 national forest parks being established within a relatively short time period of three years. On average, 70 national forest parks were designated each year during this period. This unusually rapid development of forest parks is due to the specific historical situation as aforementioned. The effects are twofold. On the one hand, a forest park system was developed in a short time period with a strong social influence. Also, its role in natural resource protection and forestry industry development was substantiated. On the other hand, forest parks were developed quickly under the circumstance that sound laws/regulations, policies and standards had not yet been formulated, leading to some places with lower quality resources to be established as national forest parks. This situation caused trouble for the management of forest parks in the following years.

1.3 The steady development stage (1994–2000)

After the three-year rapid development, momentum began to slow down as forest parks entered the steady development stage. Within seven years, a total of 110 national forest parks were designated with an annual average of less than 16 parks being established. Although national forest parks increased slowly in number during this period, the operational and administrative management of forest parks had been reinforced and a basic framework of sector administration for forest parks was formed. In January of 1994, the Ministry of Forestry enacted "Forest Park Management Regulations", and similar regulations were also promulgated in the provinces of Hunan, Sichuan, Guangdong, Shanxi, and Guizhou, among others. In December 1994, the Ministry of Forestry created the "China Forest Scenery Resource Evaluation Committee", standardized the process for national forest park establishment approval, and strengthened the technological support for this approval process. In addition, the Ministry of Forestry also enacted "Forest Park Master Plan Standards and Specifications" in 1996 (LY/T5132-95). Moreover, in 1999, the State Bureau of Quality and Technical Supervision issued "the Forest Scenery Quality Assessment of China's Forest Parks" (GB/T1805-1999), which marked an important progress in the standardization of China's forest parks. Finally, during the same time period, several forestry oriented higher institutions began to set up majors or offer courses in forest recreation/tourism.

1.4   The rapid maturity stage (2001–   )

In the five years between 2001 and 2005, a total of 283 national forest parks were approved with an annual average of more than 50. The rapid growth of forest parks during this period was due to the composite benefits of forest parks being further publicly recognized, particularly by local governments. It was also due to the rapid transition in the way that forest lands had been used as a result of a sharp decline in timber harvesting owing to the implementation of the Natural Forest Protection Project engineered by the central government. During this time period, a new paradigm, or line of thoughts on developing China's forest park system was developed under the new historical condition. In 2001, the State Forestry Administration held another national roundtable conference on forest park management. It was at this conference that the nature of the forest park cause was determined, and the new tasks and objectives for forest park development under the new era were proposed. It was pointed out at the conference that forest parks, as a cause for the public good, are an essential part of China's ecological construction and natural protection cause, an important window to showcase China's magnificent mountains and rivers, and an integral part of China's ecological and cultural construction. The conference also stressed that developing forest tourism was an important aspect of constructing the modern forestry industry system and an important way to promote an all around social and economic development of forest farms. It symbolized that a fundamental change had taken place on how to utilize China's forest resources. Developing forest tourism was an objective reflection on the shift of social economic development from one that was timber production oriented to one that was ecological demand oriented. In addition, the conference advocated to "Develop green ecology, run green businesses, and create green civilization"; to construct forest parks into typical representatives of the beautiful mountains and rivers in China; and to develop the forest tourism industry into the predominant forestry industry in the country.

2. The spatial distribution of national forest parks

2.1   The spatial distribution of national forest parks by administration

Heilongjiang province (including Longjiang Forestry Industry Enterprises and Daxinganling Forestry Industry Enterprises, hereafter the same) has 51 national forest parks, the largest number in the country. There were more than 30 national forest parks in the provinces of Shandong, Jiangxi, and Sichuan, and more than

20 national forest parks were established in provinces/autonomous regions of Zhejiang, Anhui, Liaoning, Yunnan, Hunan, Inner Mongolia (including Inner Mongolia Forestry Industry Enterprises, hereafter the same), Henan, Shaanxi, Jilin (including Jilin Forestry Industry Enterprises, hereafter the same), Hubei, Gansu, Guangdong, and Hebei. Jilin province had the largest area of national forest parks of over 1.90 million hm^2. Heilongjiang, Inner Mongolia, and Tibet each had national forest parks of over 1 million hm^2 in size.

The following patterns can be observed:

① Provinces with key state-run forest farms had a larger area of national forest parks. The top five provinces/autonomous regions with the largest areas of national forest parks are, in order of park areas, Jilin, Heilongjiang, Inner Mongolia, Tibet, and Sichuan. This reflected that forest parks depend on abundant forest resources. It also substantiated that forest parks are an important part of China's ecological construction and natural protection system.

② Provinces/autonomous regions of Jiangxi, Sichuan, Anhui, Yunnan, Hunan, Henan, and Shaanxi had a variety of high quality forest scenery resources (including biological, geological, water, climatic, and cultural resources) and have established a large number of national forest parks. This indicates that after 20 years of construction, the forest park cause has generated remarkable influence and has been accepted by governments at different levels. Those regions with plentiful and high quality forest scenery resources have been managed in conformity with the forest park management system.

③ Economically developed provinces such as Shandong, Zhejiang, Liaoning, and Guangdong were featured by forest parks as being large in number but small in size. In contrast, economically depressed provinces/autonomous regions with larger and more abundant resources represented by Tibet and Xinjiang had forest parks that were large in area but small in number. The economically developed provinces had a large tourism market base and sufficient investments in tourism development. Moreover, these provinces were more motivated to protect precious resources. Thus, local governments were more likely to place higher priority on forest-based tourism development.

2.2   Distribution of national forest parks by planning regions

According to "China's Forest Park Development Plan, 2004-2010"[3], the whole country is divided into seven regions in relation to the distribution of forest parks (Table 1).

The distribution of these seven regions by national forest park number and size. As shown, Northern China Plain and Loess Plateau has the most national

Table 1   China's Forest Park Development Planning Regions

Region	Provinces/autonomous regions
Northeastern and Inner Mongolia Mountain Plateau	Inner Mongolia, Liaoning, Jilin, Heilongjiang
Northern China Plain and Loess Plateau	Beijing, Tianjin, Hebei, Shanxi, Shandong, Henan, Shaanxi
Hilly Plain of Eastern China	Shanghai, Jiangsu, Zhejiang, Anhui, Fujian, Jiangxi
Low Hills of Central South China	Hubei, Hunan, Chongqing
Low Hills of Southern China	Guangdong, Guangxi, Hainan
High Mountains and Deep Valleys of Southwestern China	Sichuan, Guizhou, Yunnan, Tibet
High Plateau and Deserts of Northwestern China	Gansu, Ningxia, Qinghai, Xinjiang

forest parks, followed by Northeastern and Inner Mongolia Mountain Plateau, Hilly Plain of Eastern China, High Mountains and Deep Valleys of Southwestern China, Low Hills of Central South China, Low Hills of Southern China, and High Plateau and Deserts of Northwestern China.

In terms of national forest park distribution by size, the order is Northeastern and Inner Mongolia Mountain Plateau, which accounted for 45% of the total national forest park areas, followed by High Mountains and Deep Valleys of Southwestern China, High Plateau and Deserts of Northwestern China, Northern China Plain and Loess Plateau, Hilly Plain of Eastern China, Low Hills of Southern China, and Low Hills of Central South China.

2.3   Regional distribution of national forest parks

According to the national regional plan, the whole country can be divided into four great regions: Eastern China, Central China, Western China, and Northeastern China (Table 2).

The development process of China's forest parks can be divided into five time periods based on China's Five-Year Plan for National Economic and Social Development. Table 3 presents the growth pattern of national forest parks for each region. Figure 1 shows the percent that newly established national parks in each time period for each region accounted for the total newly established national parks for all of the regions. As shown, the percent for the Western region went up considerably over time, while the opposite was true for the Eastern and Central regions. Little change occurred for the Northeastern region. This indicates an obvious spatial shift in forest park distribution, moving from

Table 2   China's four great regions

Name	Provinces/autonomous regions/municipalities
Eastern	Beijing, Tianjin, Hebei, Shanghai, Jiangsu, Zhejiang, Fujian, Shandong, Guangdong, Henan (9 provinces and 1 municipality)
Central	Shanxi, Anhui, Jiangxi, Henan, Hubei, Hunan, Chongqing, Sichuan, Guizhou, Yunnan (9 provinces and 1 municipality)
Western	Inner Mongolia, Guangxi, Shaanxi, Gansu, Qinghai, Ningxia, Xinjiang, Tibet (3 provinces and 5 autonomous regions)
Northeastern	Liaoning, Jilin, Heilongjiang (3 provinces)

Table 3   Newly established national forest parks and percentages for four great regions

Year	1982~1985		1986~1990		1991~1995		1996~2000		2001~2005	
	No.	%	No.	%	No.	%	No.	%	No.	%
Eastern	3	50.0	4	40.0	60	24.6	24	28.6	65	23.0
Central	2	33.3	3	30.0	84	34.4	15	17.9	47	16.6
Western	1	16.7	1	10.0	58	23.8	34	40.5	121	42.8
Northeastern	0	0.0	2	20.0	42	17.2	11	13.0	50	17.6
Total	6	100.0	10	100.0	244	100.0	84	100.0	283	100.0

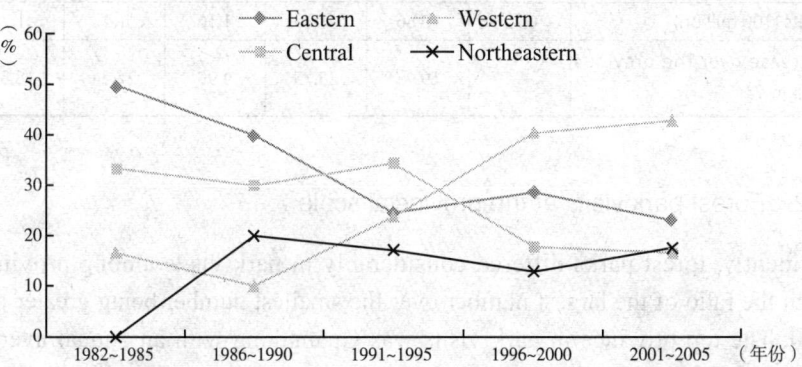

Figure 1   The percent that newly established national parks in each time period for each region

the east to the west. The occurrence of this pattern can be explained by the fact that the development of the forest tourism industry has been closely related to local economic conditions. For example, the Eastern and Central regions were characterized by good economic conditions and infrastructure with a strong capacity to develop forest tourism. As a result, these regions recognized the importance of developing forest tourism earlier than did other regions, including the Western region. However, with the gradual improvement of the economic situation in the Western region, the resource potential has been gradually discovered. As a result, more forest parks have been established in this region.

3. Forest park visits at the national scale

3.1   Change of forest park visits over the years

The change of forest park visits basically matched the development stages of forest parks. For example, during the rapid development stage, the number of forest park visits increased rapidly, while during the steady development stage, park visits remained almost the same for each year. Between 2000 and 2005, the annual growth rate was up (between 18% and 30%), except in 2003, when the low growth rate of 4.99% was due to SARS related impacts (Table 4).

Table 4   Forest park visits in the country between 2000 and 2005

Year	2000	2001	2002	2003	2004	2005
Visits (100 million)	0.72	0.86	1.10	1.16	1.47	1.74
Increase over the previous year (%)	—	19.25	28.75	4.99	27.24	18.30

3.2   Forest park visits at the provincial scale

Evidently, forest parks differed considerably in park visits among provinces with the ratio of the largest number over the smallest number being greater than 280. The top province in park visits was Guangdong with an annual average park visitation of 15 million visits, followed by Shandong and Zhejiang (over 12 million visits) and Jiangsu (over 8 million visits). The remaining provinces have each accommodated less than 6 million visits per year.

3.3   Forest park visits at the scale of planning regions

Based on the annual average park visits between 2000 and 2005, the Hilly Plain of Eastern China was visited the most, followed by Northern China Plain and Loess Plateau, Low Hills of Southern China, Low Hills of Central South China, Northeastern and Inner Mongolia Mountain Plateau, High Mountains and Deep

Valleys of Southwestern China, and High Plateau and Deserts of Northwestern China. This pattern is largely different from that of the order of the forest park size by planning regions as displayed in Figure 8. Specifically, Northeastern and Inner Mongolia Mountain Plateau was number one in national forest park areas, while the Hilly Plain of Eastern China was placed fifth (Figure 8). The opposite is true in terms of park visits; in that the former was number five while the latter was number one. This indicates that a gap existed between the two regions in their development and construction strengths, operation and management levels, and market conditions.

4. Direct tourism revenues from forest parks

4.1 Variation of forest park tourism revenues in the country

As shown in Table 5, the total direct tourism revenue from entrance fees collected from all of China's forest parks was greater than 100 million yuan in 1992, 500 million yuan in 1995, and 1 billion yuan in 2000. Since then, tourism revenues from parks have been increasing exponentially (except 2003 due to SARS related impacts) with an annual average growth rate of over 20%, and an additional earning of 1 to 2 billion yuan each year. The total revenue was up to 8.4 billion yuan in 2005.

Table 5   Total tourism revenues from all forest parks from 2000 to 2005

Year	2000	2001	2002	2003	2004	2005
Tourism revenue (100 million yuan)	12.93	28.19	37.03	41.89	69.10	83.99
Increase of percent over the previous year (%)	—	118.02	31.36	12.13	64.96	21.55

4.2 Annual average direct tourism revenues from forest parks by provinces

From 2000 to 2005, Zhejiang stood out among all provinces with an average annual tourism revenue from entrance fees of over 1 billion yuan, followed by Shandong, Jiangsu, and Guangdong.

4.3 Annual average direct tourism revenues from forest parks by planning regions

Tourism revenues earned by each planning region, to a large extent, paralleled the number of park visits received by each region. With this said, Low Hills of Southern China slightly out-numbered the Low Hills of Central South China in park visits, while the latter slightly outweighed the former in revenues from entrance fees.

The percent that tourism revenues from forest parks in each planning region accounted for the total tourism revenues from all forest parks in the country. The Hilly Plain of Eastern China and the Northern China Plain and Loess Plateau accounted for a larger proportion of total tourism revenues from all forest parks, totaling up to 58%.

4.4   Utilization benefit of forest scenery resources in each planning region

The forest scenery resource utilization benefit for a planning region was calculated by dividing the annual tourism revenues from all national forest parks in that region over the total area of national forest parks in the region. As aforementioned, national forest parks are the backbone of the forest park system, accounting for 73% of total forest parks in terms of size. In the meantime, national forest parks generated a large bulk of forest tourism benefits. Thus, this ratio can truly reflect the forest scenery resource utilization benefit for each region.

As shown in Table 6, to a large extent, the utilization benefit for each region differs from one another. Specifically, the Low Hills of Eastern China

Table 6   Ratio of annual average revenue of all national forest parks in a planning region against the total area of national forest parks in the region

Region	Total area of national forest parks ('0000 hm^2)	Annual average revenue of all forest parks('0000 yuan)	Ratio of revenue over area (yuan/hm^2)
Northeastern and Inner Mongolia Mountain Plateau	490	40270.63	82.18
Northern China Plain and Loess Plateau	98	90644.11	924.94
Hilly Plain of Eastern China	72	172493.10	2395.74
Low Hills of Central South China	47	60952.77	1296.87
Low Hills of Southern China	50	52289.83	1045.80
High Mountains and Deep Valleys of Southwestern China	214	35970.81	168.09
High Plateau and Deserts of Northwestern China	134	5589.80	41.71

has a revenue-over-size ratio of 2395.74. That is, each hm^2 of forest parks corresponded with a benefit of 2395.74 yuan. This number was only about 41.71 yuan in the High Plateau and Deserts of Northwestern China, approximately 60 times less than that in Eastern China. It is worth noting that the utilization benefit pattern generally matched the economic development level of each region, in that economically developed regions had higher ratio values, with the opposite being true for economically undeveloped regions.

5. Comparison between forest tourism and the national tourism industry

Table 7 presents forest park visits, the forest park tourism GDP and the national travel and tourism industry GDP. As indicated, the proportion of forest park tourism by the national travel and tourism industry GDP has steadily increased over the years. For example, in 2001, the total forest park visits accounted for more than 10% of total domestic tourist visits. This number increased up to 14.38% in 2005. The annual park visits during the six years (between 2000 and 2005) accounted for an average of 12.37% of total domestic tourist visits,

Table 7   Comparison between forest park tourism and the travel and tourism industry in the whole nation

Year	2000	2001	2002	2003	2004	2005	Mean
Average domestic tourism spending (yuan)	426.50	449.50	441.80	395.60	428.30	436.10	—
Total domestic visits (100 million visits)	7.44	7.84	8.78	8.70	11.00	12.12	—
Total park visits (100 million visits)	0.72	0.86	1.10	1.16	1.47	1.74	—
Percent of total park visits over total domestic visits (%)	9.65	10.92	12.56	13.31	13.39	14.38	12.37
Increase of percent in visits over the previous year (%)	—	1.27	1.64	0.75	0.08	0.99	0.95
Total travel & tourism industry GDP (100 million yuan)	4519	4995	5566	4882	6840	7686	—
Total forest park GDP (100 million yuan)	306	385	487	458	631	760	—
Percent of the total forest park GDP over the total travel & tourism industry GDP (%)	6.77	7.71	8.75	9.38	9.23	9.89	8.62
Increase of percent in GDP over the previous year (%)	—	0.94	1.04	0.63	-0.15	0.66	0.62

an annual increase of 0.95%. In addition, in 2005 the total forest park tourism GDP accounted for approximately 10% of the total national travel and tourism industry GDP. The annual GDP from forest parks accounted for an average of 8.62% of the total national travel and tourism industry GDP during the same period with an annual increase of 0.62%.

6. Discussion and conclusion

6.1   Obvious development patterns of China's forest parks and forest tourism

Forest parks have undergone four development stages which are featured by a repeating pattern that alternately occurred between rapid development and steady development, which were subject to external factors such as policy and market. Temporally, forest parks developed quickly over time. Spatially, the distribution of forest parks shifted from the eastern region to the western region. Additionally, forest park visits corresponded with forest revenues. Finally, forest parks played an increasingly important role in the national tourism industry.

6.2   Rapid forest park development

With the rapid social and economic development, the increasing recognition of the importance of forest park tourism, and more non-state owned forest resources being set aside as forest parks, forest parks will increase rapidly in number during the Eleventh Five-Year Plan period. According to "China's Forest Park Development Plan, 2004-2010," the number of forest parks in the country will jump from the current 1928 to 2800 in 2010.

6.3   Increasing importance of forest tourism in western China

During the Eleventh Five-Year Plan period, forest parks in western China will account for an increasing proportion of total forest parks in number, as will forest tourism revenues. With more landscape resources being developed and utilized for tourism in this region, more forest parks will be established in the region during this time period. Thus, spatial shifts in forest park distribution will continue to exist. As shown in Table 8, forest tourism revenues for Chongqing, Sichuan and Yunnan were greater than the national averages.

Table 8   Increasing patterns of forest tourism revenues for several western provinces (2001-2005)

Year	2001	2002	2003	2004	2005
Chongqing ('0000 yuan)	7404.23	8439.30	12813.54	36861.02	52184.59
Increase of percent over the previous year (%)	—	13.98	51.83	187.67	41.57
Sichuan ('0000 yuan)	12467.62	7084.84	17066.90	52113.61	84102.99
Increase of percent over the previous year (%)	—	-75.98	140.89	205.35	61.38
Yunnan ('0000 yuan)	563.00	798.31	2874.59	4006.17	6480.89
Increase of percent over the previous year (%)	—	41.80	260.08	39.36	61.77
Increase of percent over the previous year for the whole country	—	31.36	12.13	64.96	21.55

6.4   Increasing forest park visits and forest tourism revenues

During the Eleventh Five-Year Plan period, forest park visits and forest tourism revenues are predicted to increase at an annual rate of 20%, with 400 million park visits and over 20 billion yuan projected for 2010 (Table 9).

Table 9   Predicted forest park visits and direct forest tourism revenues (2006 – 2010)

Year	2006	2007	2008	2009	2010
Forest park visits (100 million visits)	2.09	2.51	3.01	3.61	4.33
Direct forest tourism revenues (100 million yuan)	100.79	120.95	145.13	174.16	208.99

6.5   Increasing importance of forest tourism to national travel and tourism industry

Given that forest tourism is increasingly attractive among the public, it is estimated that annual park visits and the forest tourism GDP will account for 19% of total national domestic visits and 13% of the national travel and tourism industry GDP by 2010 (Table 10) during the Eleventh Five-Year Plan period. There are no signs that this trend will cease in the near future.

Table 10   Predicted proportion of forest park visits and forest tourism (2006 – 2010)

Year	2006	2007	2008	2009	2010
Percent of park visits by total domestic visits (%)	15.33	16.28	17.23	18.18	19.13
Percent of forest park tourism GDP by the national travel and tourism GDP (%)	10.51	11.13	11.75	12.27	12.89

Reference

[1] The editing commission of China forestry yearbooks. *China Forestry Yearbooks 1982–2005* [M]. Chinese Forestry Press, 1983–2006 (*in Chinese*) 中国林业年鉴编纂委员会:《中国林业年鉴 (1982~2005)》[M], 中国林业出版社, 1983~2006。

[2] The Forest Park Office of the State Forestry Bureau. *Forest park construction and development–China forest park roundtable conference proceedings* [M]. The Forest Park Office of the State Forestry Bureau, November, 2001(*in Chinese*) 国家林业局森林公园管理办公室:《森林公园建设与发展——全国森林公园工作会议典型材料》[M], 国家林业局森林公园管理办公室, 2001。

[3] The State Forestry Bureau. *China Forest Park Development Plan (2004–2010)* [M]. The State Forestry Bureau, July, 2004(*in Chinese*) 国家林业局:《全国森林公园建设规划 (2004~2010)》[M], 国家林业局, 2004。

Pareto Optimum Condition of the Tourism Marketplace Promotion and Evaluation of Implement Mode

Yu Jie[1], Chen Rong[2], Li Shumin[1]

(1.School of Economics & Management, Northwest University, Xi'an 710069;
2.CITS (Group) Corp, Beijing 100005)

Abstract: The paper explains the tourism firms' behavior on the tourism marketplace promotion with the help of the theory of Pareto optimum condition of the pure public product. Applied the efficient condition of the supply of the tourism marketplace promotion to evaluate the mode of the government oriented and the private sector oriented, the paper holds that the two modes offering the tourism marketplace promotion can lead to Pareto inefficiency, but the combined mode of the government and the private sector can implement Pareto improvement. Based on the combined mode, a new management structure of the tourism marketplace promotion can be designed. The paper also suggests that tourism association should play an efficient role in the tourism marketplace promotion.

Key words: tourism marketplace promotion; public product; Pareto Optimum; tourism association

[About authors] Mrs. Yu jie, Ph. D. of economics, lecturer of tourism management department in the School of Economics and Management ,Northwest University, with research interests centered on issues related to tourism economics; Mr. Chen Rong, vice president of China International Travel Service Group, doctor of Management Science and Engineering, Xi'an Jiaotong University, mainly concentrated on management of tourism corporation; Mr. Li Shumin, tutor of doctor of tourism management, dean of tourism management in School of Economics and Management , Northwest University.

As for the promotion subject in the current tourism marketplace of our country, there are five kinds: cooperation promotion of interregional governments, cooperation promotion of interregional tourism enterprises, cooperation promotion between governments and tourism enterprises, promotion of regional government and promotion of large-scale tourism enterprise. The above mentioned five kinds can be concluded three coordination modes: the government oriented, the private sector oriented and the combination of the government and the private sector. In the promotion games of the governments and enterprises, tourism promotion expenses mainly come from the regional public financial allocation, the tourism publicity expense paid by tourism enterprises and the tourism development fund paid by tourism department. Among them, the public financial allocation of the government, which is the chief source of tourism promotion expenses currently, accounts for 63%, with the tourism publicity expense 14%, the tourism development fund 6%,and other sources(such as cumulative surplus over the years) 17%. In the promotion of single subject, it is difficult for governments and enterprises to collect a huge sum of promotion expenses due to the pressure of the public financial budget. In the cooperation promotion, the activities are ultimately unsuccessful owing to the reasonable behavior as"hitch a ride" of the subject and the independent gambling between the governments and the enterprises. The supply inefficiency of the tourism marketplace promotion directly affects the promotion of the whole image about the tourism destination and the marketing of the tourism enterprise.

1. The economics property of tourism marketplace promotion

1.1   Tourism marketplace promotion is a public product

Because of the tourism behavior of purchase and consumption occurring in different places and the difference of tourism experience, the promotion of tourism products for the tourist source markets have a direct effect on the purchase decision of tourists. The consumption of tourists in the destinations includes food, accommodation, transportation, touring, purchasing and entertainment. In the view of the welfare economics, this kind of consumption can be regarded as the various distribution of merchandise with a fixed quantity from tourist source market among several economics subject in tourism destination. Therefore, the potential of the source market brought by organizing the promotion will benefit for the whole tourism industry chains in the tourism destination. In view of this point, tourism marketplace promotion is a public product.

1.2   From the point of industry, tourism marketplace promotion is a purely public product

Once masterminding the image of the tour destination, packaging and marketing in tourist source market and latent visitors becoming realistic visitors, the industry chains in the destination, including the travel agency, the hotel, the scenic spots, tourist transportation and souvenir shops, share the same tourists and gain the corresponding benefits equally. Therefore, from the point of the industry, tourism marketplace promotion, which is a purely public product, meet the condition of the characteristics of non-competition and non-exclusion.

1.3   From the point of enterprise, tourism marketplace promotion is a semipublic product

As a result of unequal distribution of tourists among different enterprises in the same industry of the destination, it's difficult to exclude the share that some enterprises don't pay for the image promotion of the destination through price mechanism. Consequently, from the point of single enterprise, tourism marketplace promotion is viewed as a semi-public product or a hybrid product and service which are intervenient between purely public product and purely private product owing to its trait of non-exclusion as a public product and trait of competition in consumption.

2. The theory basis of the behavior analysis on tourism marketplace promotion

In the traditional theory of public product, public product is studied as the ultimate consumer goods or the middle consumer goods. The assumption of analysis is that the purely public goods are provided in the way of imposing one-off total amount tax (the consistency pricing) by the government. In this frame of the theory, purely public product should be provided by government, and all families as the main economic subject consume the public goods collectively. However, different consumers hold different comments on provision of public goods while all consumers need to pay the same price for them. This makes the competition balance lose efficiency, which means governments would play a certain role in provision public goods and correction the malfunction of market. In the process of trying to gain an efficient result, the biggest restriction of governments is to understand family inclination and payment desire for public goods. Linder Benefit Pricing method (Characteristic Pricing Method), which is designed for this purpose, can't also eliminate the phenomenon of "hitch the ride" caused by positive exteriority of purely public goods. While the public goods is completely provided voluntarily by the family, everybody can pay a price that reflects personal evaluation. Deducting the balance of private supply in the economy of Linder balance, Pareto improvement and shortage of public goods appear [1]. Therefore, it's impossible to both solve the problem of "hitch a ride" and guarantee Pareto Optimum at the same time in the process of designing supply mechanism for public goods.

Although the premise assumption of this conclusion is that family as the main economic subject consuming the public goods and deduction the Pareto Optimum condition for purely public goods by the way of utility function that can show preference, it also can be used to explain the economic behavior of the tourism enterprises in the tourism marketplace promotion. As for the promotion in tourist source market, travel agencies combine the scenic spots, hotels, transportation and shopping to meet different needs of tourists. Therefore, the tourism promotion for the same tourist source market of the enterprises in the two sectors of travel agency and scenic spots which is carried out for the purpose of increasing the amount of tourists is a purely public product. At this time, these two sectors exist as the consumers rather than the general manufacturers. Then it's beyond doubt that we can use the consumer demand theory of maximum utility in the public economic theory to explain the condition of maximizing subject in the tourism marketplace promotion, that is to say, in the realistic economy whether there is a condition that makes both scenic spots and travel agency would like to conduct the promotion for the same tourist source market. This is a matter of Pareto efficiency which obeys Samuelson rules and furthermore whether an implement mode can make this cooperation effective.

3. The condition of pareto optimum in the tourism marketplace promotion

Supposed that there are two sectors in the same tourism destination, Sector x_1 stands for the section of scenic spots, while Sector x_2 stands for the section of travel agency. The two sectors are trying to make a decision about whether conducting promotion for the same tourist market. The utility function of Sector x_1 is showed by $u_1(x_1, G)$ and the utility function of Sector x_2 is showed by $u_2(x_2, G)$, in which x_1 means funds of their own and G means the public goods in tourism marketplace promotion. Here, G can be 0, which means that promotion is not carried out, while 1 means that promotion is conducted.

In order to deduce the condition of Pareto Optimum, firstly, the allocation of resources of Sector x_1 should be the maximized utility in the condition that the utility of Sector x_2 keep fixed as $\bar{u}_2$. Then constructing the Lagrange Function:

$$z=u_1(x_1,G)+\lambda_1 \left[u_2(x_2, G)-\bar{u}_2 \right] + \lambda_2 \left[F(x_1+x_2, G) \right]$$

making the first order condition:

$$\left[\partial u_1/ \partial G \right] / \left[\partial u_1/ \partial x_1 \right] + \left[\partial u_2/ \partial G \right] / \left[\partial u_2/ \partial x_2 \right] = F_2/ F_1$$

The equation shows that the Pareto Optimum condition of the two sectors for promotion synchronously is that the summation of separate devotion of these two sectors for promotion and the marginal substitute rate (mrs) of surplus fund equals to the marginal exchange rate (mrt) of summation of the total promotion devotion and total surplus fund of two sections:

$$mrs_1+mrs_2=mrt(G,X)$$

In an economy where uses price as accounting unit, this condition of optimum can be illustrated by marginal benefit (mb) and marginal cost (mc).Moreover, it can be expanded to the Pareto Optimum condition of tourism marketplace promotion involving many sectors on the base of this basic mode.

$$\sum_{i=1}^{1} mb_i(G) = mc(G)$$

This formula indicates that the optimal utility condition of conducting tourism marketplace promotion which is viewed as purely public goods is the sum of marginal benefit for all sections equals to marginal cost all sections pay for tourism marketplace promotion.

4. The implement mode and evaluation of tourism marketplace promotion

4.1 The mode of the government oriented

The demand curve of tourism marketplace promotion as public product (G) can be depicted as total demands of all sections' preference to buy a certain quantity of public goods at different price levels. Because of the different preferences and levels of wealth, each section sets price according to its own benefit for a fixed G. As Figure1 shows, the demand curve D_1 of Section x_1 indicates that it has a lower demand for public goods G_1 in the condition that government imposes an unified promotion fee, and just like the world heritages which enjoy a certain reputation,such as the Summer Palace and Terra Cotta Warriors and Horses, their images have already been regarded as a symbol of a region even a country, so these units would like to pay a lower cost for public goods. D_2 needs more public goods, and just like travel agencies and hotels which can gain a wide range of brand identity through the tourism marketplace promotion, therefore they would like to pay a higher cost for public goods. However, the traitor of public goods determines that the consumption of D_1 and D_2 is the total consumption of public goods.

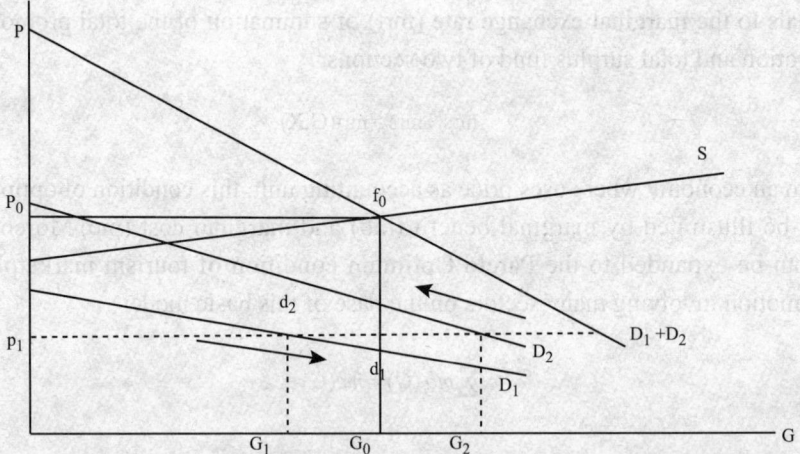

Figure 1   The Linder Balance of Tourism Marketplace Promotion

According to a repeated operation principle by Linder, governments increase the promotion fee of Section X_2, while reducing that of Section X_1. Therefore, X_2's demand for tourism marketplace promotion will decrease along the demand curve D2 with the price tending to the point d_2, while X_1's demand for tourism marketplace promotion will increase along the demand curve D_1 with the price tending to the point d_1.This process is repeated several times until the sum of price ($G_0d_1+G_0d_2$) both parts would like to pay for a unit of public product equals to marginal benefit (G_0f_0) at the realistic supply level. At that time, both parts need an equal quantity of public goods(G_0) and the sum of benefit price for two sections equals to their marginal costs with G staying at its optimal point G_0.

The supply of tourism marketplace promotion satisfies the condition of Pareto
Optimum, and realizes Linder Balance [2].

If the cooperation of two sectors can reflect their evaluation of tourism
marketplace promotion, it is easy for them to decide whether they should get together
to promote. According to the conclusion in the famous Samuelson's dissertation
titled Pure Theory of Public Expenditures, the rational people wouldn't like to
expose their true preference for public goods, which can also be explained as
their marginal cost or marginal benefit [3]. Because of the reliability problem
of the preference manifestation,the section probably conceals its true marginal
benefit from tourism marketplace.When sections reveal desires incorrectly in
order to adjust the equilibrium price for their own profits, the strategy damages
the foundation of Linder Balance and also lead to inefficiency of the allocation
of tourism marketplace promotion by price mechanism as public goods.

4.2   The mode of the private sector oriented

In an economy where promotion of two tourism sectors exist, the supply
equilibrium of private sectors occurs the point where indifference curves of
these two sections intersect. Section X_1 and Section X_2 satisfied the condition as
followed:

$$U_G^i/U_x^i=1(i=1, 2), \text{ hereinto}, U_G^i= \partial U^i/ \partial G, U_x^i= \partial U^i/ \partial x$$

The solution to the equation is decided by inverse functions of $g_1=\rho_1(g_2)$
and $g_2=\rho_2(g_1)$. Inverse functions of two sections are satisfied at the point S of
intersection, then the equilibrium of private supply is achieved. The point S is
Pareto Optimum when two indifference curves of these two sections tangency
meet at point S. The total optimal point of Pareto social welfare function must
lie in any point on the tracks of the effective allocation set of Pareto (contract
curve). However, any point on the contract curve is higher than the point S. That
is to say, the total quantity of supply is smaller than the optimal quantity at this
point. Compared with the optimal allocation of Pareto preference, the supply
equilibrium of private sector would lead to supply shortage of public goods.

In the condition of cooperation, the production of public goods will increase.
Figure 2 illustrates that the increase of promotion expenses paid by two sectors
aiming at the same market leads to Pareto improvement. But it's difficult to reach
the objective of high-effective equilibrium with the increase of the participants in
the tourism marketplace promotion. Because the whole benefits from promotion
exceed its cost, the supply shortage of public goods exists in any time.

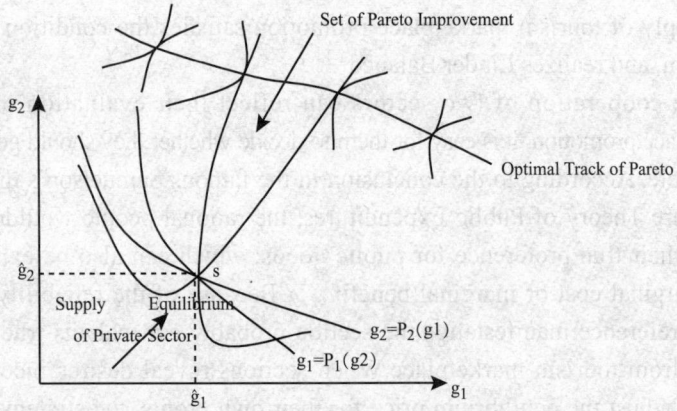

Figure 2   Non-optimal Trait of Tourism Marketplace Promotion Provided by Private Sectors

4.3   The combined mode of the government and the private sector

Is there a possibility to eliminate the problem of "hitch a ride" completely, and at the same time ensure Pareto optimum, the balance of budget and incentive compatibility? Hurwicz (1979) and Laffont (1987) argued that the possibility didn't exist at all, and it's better to give up the request of Pareto Optimum. Also the incentive compatibility should be regarded as a redundant constraint and viewed "hitch a ride" as a hypo-optimal problem. In the condition of satisfying Pareto hypo-optimum in the tourism marketplace promotion, we hope to rely on an association, like strengthening the cooperation of government, to solve the problem that inherent non-cooperation of voluntary supply in promotion, to change non-cooperation into cooperation and ultimately to gain Pareto improvement of tourism marketplace promotion.

While analyzing the relations between governments and sectors both participating in the combine promotion and not participating in it, we assume that governments maximize their preference under fixed restricted conditions, governments and tourism sectors are both sides in a gamble, and the strategy of government is providing subsidy or not while the strategy of tour sectors is attending or not.

Table 1 shows matrixes of governments and tourism sectors regarding to different strategies mix.

With reference to each couple of numbers shown in the table, the former means the benefits of tourism sectors and the latter is the benefits of governments.

c_1=Cost of tourism sectors for attending promotion (promotion fee);

c_2=Opportunity cost of tourism sectors for not attending promotion

Table 1   The Gamble Matrixes of Governments and Tourism Sectors Regarding to the Promotion Strategies

		Government	
		Subsidizing（γ）	Not subsidzing（1-γ）
Tour sector	Attending（θ）	$-c_1+k,-k$	$-c_1, 0$
	Not attending（1-θ）	$-c_2, c_2+k$	$-c_2, 0$

(information fee,etc.);

k=The subsidy of governments to sectors involved in promotion;

θ=The probability of tourism sectors to attend promotion;

γ=The probability of providing subsidy by governments to sectors involved in promotion;

The expecting benefits of governments $E_G(θ, γ)=γ(-θk+(1-θ)(c_2+k)+(1-γ)(θ0+(1-θ) 0)$;

The expecting benefits of tourism sections $E_P(θ, γ)=θ(γ(-c_1+k)+(1-γ)(-c1))+(1-θ)(-c_2γ+(1-γ)(-c_2))$;

The according first order condition: $∂EG/∂γ=θ(-c_2-2k)+(k+ c_2)$

$∂E_P/∂θ=γk-c_1+c_2$

The solution: $γ^* = \dfrac{c_1-c_2}{k}, θ^* = \dfrac{k+c_2}{2k+c_2}$

At this time, governments and tourism sectors reach the equilibrium. According to the probability θ* of tourism sectors to attend promotion, when $\dfrac{c_2}{k}$ is increasing, in other words, when opportunity cost c_2 of tourism sectors for not attending promotion is increasing, the probability of tourism sectors to attend promotion is increasing accordingly. Therefore, when governments provide some subsidies or implement some protection measures, the non-cooperation can be banned and cooperation promotion of tour sectors can be encouraged.

4.4   Evaluation of the three modes

Pricing for the promotion of tourism marketplace by mode of the government oriented, all sectors aren't likely to show their true evaluation on promotion, because they know the share of cost they are responsible for in promotion is decided by the amount of relevant marginal benefits. Then the motivation of reducing their payable share of cost by revealing lower marginal benefits than realistic one will appear and the sector to hitch a ride will come up.

In the mode of the private sector oriented, sectors hold a psychology of speculation: if other sectors attend the promotion without their involvement, this sector can get more benefits than that got from all parties'attending the promotion, therefore all sectors generate a motivation of not attending the promotion. When there isn't cooperation, all participants face "prisoner puzzle" and attempt to invest independently in promotion to gain a larger market share at the cost of other sectors'loss. However, non-cooperation leads to a unilateral promotion of tourism destination's image, the repeated construction of projects, a serious waste of fund and finally damages the total benefits of all investors and generate inefficiency of tourism marketplace promotion.

However, when promotion are carried out frequently, the decision-making process can be seen as a repeated gamble of private sectors about whether they should attend promotion or not. Supposed that there is an agreement that all parties should attend the promotion among sectors, and when anyone breaks the agreement, other parties would take a non-cooperation attitude in the next gamble. For example, when cooperation is needed in other aspect (like communication of tourism information), other parties aren't willing to work with the party who breaks the agreement. When net benefits of cooperation required by party who doesn't attend promotion outweigh that gained at the cost of benefits losses of other parties involved in promotion, this uninvolved party would choose attending rather than not and will stick to the same law in the future.

In the combined mode of the government and the private sector, on the one hand, governments try to increase the quantity of public goods offered by private sectors through providing some subsidy, in other words, to increase the participants of promotion, and which can make the equilibrium point of private sector's supply to near the track of Pareto Optimum as much as possible; on the other hand, increasing opportunity cost of non-involvement by means of changing tourism marketplace promotion from purely public product to club product through charging club member fee with the help of intermediary organization. In this way, ride hitcher can be excluded from beneficiaries, non-club members need to pay high information fee and cooperation cost if they want to share the benefits of promotion. As a result, marginal benefit of uninvolved party will be smaller than its marginal cost. Therefore, the problem of tourism sector revealing a true preferences at a lower level and non-cooperation strategies can be solved by government's incentive mechanism of subsidy and action restriction among participants, and the participants would find that lying strategy just do harm to their profits so that they expect to establish a cooperation mechanism to show their true preferences.

5. Conclusion and suggestions

Through analysis about the efficient standard level of supply for tourism marketplace and comparing two economy equilibrium achieved separately by the mode of the private sector oriented and government oriented, we find out that it's impossible to reach the Pareto Optimum condition in the two modes because there are "hitching a ride" and Pareto inefficiency of shortage of public good. However, in the combined mode of the government and the private sector Pareto improvement can be achieved, that's to say, there is the rearrangement which can make the promotion supported by governments successful and encourage private sectors to attend cooperative promotion.

Just as Stiglitz emphasizes: "Don't make government and market stand facing each other, instead, we should keep a proper balance between both sides, because of the existence of many organizations of middle economic form, such as local governments and co-operatives."The policy suggestion to Chinese government made by Jean-Claude Baumgartner, the CEO of the World Tourism and Travel Committee(WTTC) shows that China should distribute the marketing and promotion functions of National Tour Administration to tourism association that is established as an administrative institution involved both public and private sectors and its operation fund should be provided by governments. The tourism association can reduce the budget pressure of government, reduce the outlay pressure of tourism enterprises, change the inefficient operation such as the insufficient information of public decision and neglecting the analysis of cost and benefit resulted from the public goods being offered by the government alone. Therefore, tourism association, as non-government organization apart from market and government, can offer service to gain cooperation of tourism promotion subject in a more efficient way.

In order to construct the combined mode of the government and the private sector, we can establish a new organizational structure of tourism marketplace promotion with the attendance of tourism association (Table 3)

In the organizational structure, tourist association, as a key connecting government with market, is a folk organization separating from government in an autonomic position, not an organization attaching to government and not voluntary activities organized by means of administration①.It get funds from

---

①   The existing tourism associations in our country is attaching to government essentially, relying on government badly or being controlled by government, and don't provide enterprises with services of club such as arbitration of industrial dissension, establishment of industrial standard, communication of information, inspecting the dealing order of market, maintaining the profits of club members.

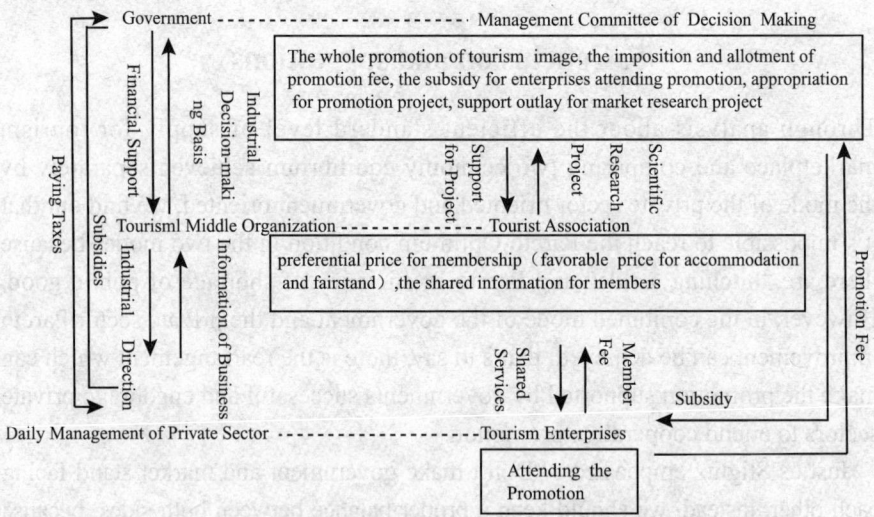

Table 3   New Organizational Structure of Tourism Marketplace Promotion

government, voluntary contribution of individuals or institutions, membership dues and operation by means of commerce and socialization. As for the self-support methods, organizing great marketing promotion, holding business lectures and symposiums, offering information and lawful advice should be paid. All participants of the activities should pay for them, but the entry fee of non-membership enterprises would be higher than that of membership enterprises. Therefore, tourist association has double commercial status: one is related to offering services for membership enterprises, and another is related to self-operation. The commercial operations of industrial association can get rid of excessive dependence on government, and gain the really organizational autonomy on personnel arrangement and association function. It's possible to predict that tourist association will take more responsibilities of offering public service on tourism marketplace promotion in the future.

References

[1] Robin W. Boadway, David E. Wildasin. Public Sector Economics [M]. Second Edition. Beijing: China Renmin University Press, 2000: 60~93 (in Chinese). 〔美〕鲍德威、威迪逊:《公共部门经济学（第二版）》[M]，邓力平主译，北京，中国人民大学出版社，2000年，第60~93页。

[2] Gareth D.Myles. Public Economics [M]. Beijing: China Renmin University Press, 2001:

248–292 (in Chinese). 加雷斯 · D.迈尔斯:《公共经济学》[M], 匡小平译, 北京, 中国人民大学出版社, 2001 年, 第 248~292 页。

[3] Samuelson P. A. Pure theory of public expenditures [J] . Review of Economics and Statistics, 1954, (36): 387–389.

[4] Raghbendra Jha. Modern Public Economics [M] . Beijing: China Youth Press, 2004: 94 –130 (in Chinese).〔英〕拉本德拉 · 贾:《现代公共经济学》[M], 王浦劬、方敏等译, 北京, 中国青年出版社, 2004 年, 第 94–130 页。

[5] Hurwicz L. Outcome functions yielding Walrasian and Lindahl allocations at Nash equilibrium points [J] . *Review of Economic Studies*, 1979, (46): 217–225.

[6] Laffont J.*Handbook of Public Economics*[M] . Amsterdam: North Holland, 1987: 537–569.

[7] Qian Chunxian. WTTC publish the report on Tourism in China. [EB/OL] . http://www. xinhuanet.com, 2003–10–14(in Chinese). 钱春弦:《世界旅游及旅行理事会公布中国旅游研究报告》[EB/OL] . http://www.xinhuanet.com, 2003–10–14。

Toward Tourism Cooperation among 4 Destinations across the Taiwan Straits: A Marketing Perspective

Li Tianyuan

(Business School, Nankai University, Tianjin 300071, China)

Abstract: By reviewing tourism to Chinese mainland, Hong Kong, Macau and Taiwan respectively, the position of these destinations as a whole in Northeast Asia and even in Asia Pacific tourism, the common feature of their source markets, the scenario and potential in providing business among themselves, and the importance for them to establish tourism cooperation are analyzed. Based on these analyses, the realistic strategies for them to develop cooperative marketing are discussed.

Key words: inbound tourism; the Taiwan straits; cooperative marketing; tourist market

1. Some connected analyses

1.1 Asia pacific's position in global tourism

According to the World Tourism Organization and Pacific Asia Travel Association (PATA), the volume of tourist arrivals in global tourism has kept growing since mid 1990s except in 2003 the annual growth rate had a slight fall (-1.2%). During the period between 1996 and 2004, the average growth rate of

[About author] Mr.Li Tianyuan, professor of tourism in the Business School at Nankai University, with research interests centred on issues related to the theoretical studies of tourism and marketing in travel and tourism.

World tourism reached 3.1%, and the tourist arrivals to Asia Pacific region increased by nearly 6.3%, it is next to that of Middle East and ranks the second in the world. It is worthy of notice that despite the huge impact of SARS in 2003, international tourist arrivals worldwide in 2004 increased by 10.6%, with Asia Pacific leading high at 27.8%. According to the statistics, Asia Pacific region hosted 153 millions of international visitors in 2004, or 20% of the market share in global inbound tourism (Table 1). It is quite clear that Asia Pacific has exceeded America as far as the volume of international tourist arrivals is concerned, ranking the 2nd place in the pattern of world tourism compared with its original 3rd place in the past. And this change marked an unprecedented breakthrough that Asia Pacific has made in developing tourism.

Table 1   International Tourist Arrivals by Region (mil) 1996-2004

	1996 (%)	2000 (%)	2003 (%)	2004(%)	AAGR (%) 1996-2004	Y-o-Y%Change 2003-2004
World	599.6 (100)	687.3 (100)	690.0 (100)	763.0 (100)	3.1	10.6
Africa	21.9 (3.7)	27.4 (4.0)	30.8 (4.5)	33.2 (4.4)	5.3	7.8
America	116.9 (19.5)	128.0 (18.6)	113.1 (16.4)	125.8 (16.5)	0.9	11.2
Asia Pacific	93.4 (15.6)	115.3 (16.8)	119.3 (17.3)	152.5 (20.0)	6.3	27.8
Europe	353.3 (59.0)	392.7 (57.1)	396.6 (57.5)	416.4 (54.6)	2.1	5.0
Middle East	14.1 (2.4)	24.0 (3.5)	30.0 (4.3)	35.4 (4.6)	12.2	18.0

Source: Adapted from WTO and PATA

Note: The numbers of Asia Pacific in the table excludes North America and Chile.

AAGR is Average Annual Growth Rate.

1.2   Northeast asia's share of asia pacific tourism

According to PATA, the total tourist arrivals to Asia Pacific countries including North America and Chile exceeded 306 millions in 2004, and 285 millions or 93.1% of them were received by 41 countries and areas in the region.[1] By analyzing the related data, it can be found that Northeast Asia's share of Asia Pacific tourism accounts as high as 56.96% (Table 2). This shows to a great extent that Northeast Asia occupies the most important position in Asia Pacific tourism as a whole. If omitting those arrivals to North America and Chile in order to make an easier understanding, the role that Northeast Asia plays in Asia Pacific tourism can be seen more clearly.

Table 2   Overseas Arrivals to Asia Pacific by Destination Area in 2004

Destination Areas	Arrivals (mil)	Market share %
Total Arrivals to Asia Pacific	285.17	100.00
North America and Chile	59.10	20.72
Northeast Asia	162.43	56.96
Southeast Asia	48.40	16.97
South Asia	5.86	2.05
Oceania	9.33	3.27

Source: Adapted from *Asia Pacific Tourism Forecasts 2006-2008*

1.3   Position of the destination area with Chinese mainland, Taiwan, Hong Kong and Macau as a whole in northeast asia tourism

Like the role Northeast Asia plays in Asia Pacific tourism, the destination area with Chinese mainland, Taiwan, Hong Kong and Macau as a whole holds an important position in Northeast Asia tourism. This can be seen clearly in Table 3. For instance, the total overseas arrivals to Northeast Asia in 2004 were 162433171, with 109038218 of them to Chinese mainland, 2950342 to Taiwan, 21810630 to Hong Kong and 16672556 to Macau. In other words, if these are added together, the total overseas arrivals to the destination area with Chinese mainland, Taiwan, Hong Kong and Macau as a whole would reach 150.47 million, or 92.64% of total tourist arrivals in Northeast Asia.

With the above observation and analyses, some conclusions can be made as

Table 3   Overseas Arrivals to Chinese mainland, Taiwan, Hong Kong and Macau as a Share of Total Arrivals to Northeast Asia 2004

Destinations	Arrivals (10 thousand)	Market share %
Northeast Asia	16243.32	100.00
Chinese Mainland	10903.82	67.13
Taiwan	295.03	1.82
Hong Kong	2181.06	13.43
Macau	1667.26	10.26
（Sub-total）	(15047.17)	（92.64）
Japan	613.79	3.78
South Korea	551.84	3.40
Mongolia	30.51	0.19
North Korea	na	na

Source: Adapted from Asia Pacific Tourism Forecasts 2006-2008

following:

• For half a century after World War II, more than 90% of the tourist arrivals worldwide were shared orderly by Europe, America and Asia Pacific like 3 legs of a tripod. With the rapid growth of tourist arrivals in the last decade, however, the development of Asia Pacific tourism has brought a change to the traditional pattern of world tourism, and Asia Pacific has come to the 2nd place instead of its 3rd place in the past.

• The combined market share of Chinese mainland, Taiwan, Hong Kong and Macau as a whole has become so big that it reflects not only this area's position in Northeast Asia or even in Asia Pacific tourism, but also suggests the area's huge potential of tourism growth.

2. Main source markets of the destination area across the taiwan straits

2.1   Main source markets of tourism to Chinese mainland

Hong Kong, Macau and Taiwan have been in fact parts of China, having close relations with the inland geographically, culturally and ethnically. Hong Kong, Macau and Taiwan all enjoy a higher level of economic development, with Hong Kong and Taiwan sharing for a long time the title of economic 'Little Dragon' in Asia. With the deepening of China's policy of reform and open-door to the outside world and large number of businessmen coming from Hong Kong, Macau and Taiwan to make investments in the mainland, the exchange of visits between these destinations has been increasing. Politically, both Hong Kong and Macau are now Special Administrative Regions of China. With one China principle as the prerequisite, it is becoming a common understanding for both sides across the Taiwan straits to maintain the status quo of peace and development. With the effect of all these factors, Hong Kong, Macau and Taiwan as a whole has been the largest source market for the mainland tourism industry, and will continue to be its stable source market in the days to come. [2]

For instance, of the 109038218 overseas arrivals to Chinese mainland in 2004, those from Hong Kong, Macau and Taiwan represent 85%, and those arrivals from Hong Kong, in particular, represent as high as 61%. [1]

As far as the foreign arrivals to Chinese mainland are concerned, the top 10 source markets in 2004 consisted of Japan, South Korea, Russia, USA, Malaysia, Singapore, Mongolia, Philippines, Thailand and UK. Among them, Japan had the highest market share. Of the foreign arrivals to Chinese mainland in 2004,

19.7% of them came from Japan. However, according to PATA, South Korea is projected to exceed Japan in terms of tourist arrivals to China and will become the largest source market by 2008, while the ranking pattern of the other top 8 source markets would generally remain unchanged (See Table 4).

Table 4 China's Top 10 Source Markets (Foreign Arrivals) and Their Market Share (%)

Source Market	2000	2001	2002	2003	2004	2008（forecasts）
Japan	21.7	21.3	21.8	19.8	19.7	17.3
South Korea	13.2	15.0	15.8	17.1	16.8	19.7
Russia	10.6	10.7	9.5	12.1	10.6	10.3
USA	8.8	8.5	8.3	7.2	7.7	7.4
Malaysia	4.3	4.2	4.4	3.8	4.4	5.1
Singapore	3.9	3.7	3.7	3.3	3.8	3.6
Mongolia	3.9	3.4	3.4	3.7	3.3	3.0
Philippines	3.6	3.6	3.8	4.0	3.2	2.9
Thailand	3.8	2.7	2.9	2.4	2.7	2.9
UK	2.8	2.7	2.6	2.5	2.5	2.2

Source: PATA 2006

2.2 Main source markets of tourism to Taiwan

The composition of the main source markets for inbound tourism to Taiwan in the past decade remained generally unchanged, with the order of the top 10 source markets as Japan, Hong Kong, USA, South Korea, Singapore, Thailand, Malaysia, Philippines, Indonesia and Canada. While some slight changes occasionally occurred in their ranking pattern (See Table 5), Hong Kong has remained the 2nd place. It is worthy to notice that Chinese mainland has not been listed as a source market of Taiwan because of the policy of Taiwan authority. However, should a breakthrough be made in its policy, it could be expected that Chinese mainland would soon become the largest source market of Taiwan, and a great change would therefore take place in the pattern of its source markets.

2.3 Main Source Markets of Tourism to Hong Kong

Chinese mainland, Taiwan and Macau have all been the main source markets of Hong Kong tourism for a long time. And this situation will continue according

Table 5　Taiwan's Top 10 Source Markets and Their Market Share (%)

Source Market	2000	2001	2002	2003	2004
Japan	34.9	37.1	36.2	29.2	30.1
Hong Kong	13.8	15.0	16.0	14.4	14.1
USA	13.7	13.0	13.0	12.1	13.0
South Korea	3.2	3.2	2.9	4.1	5.0
Singapore	3.6	3.7	3.9	3.5	4.0
Thailand	5.1	4.4	3.9	4.4	3.5
Malaysia	2.2	2.2	2.4	3.0	3.1
Philippines	3.2	2.6	2.7	3.6	2.9
Canada	1.5	1.5	1.6	1.5	1.7
Indonesia	4.1	3.4	3.2	1.7	1.5

Source： PATA 2006

to the related forecasts made by PATA. In addition, those also included in Hong Kong's top 10 source markets are Japan, USA, South Korea, Singapore, UK, Australia and Malaysia (See Table 6).

Table 6　Hong Kong's Top 10 Source Markets and Their Market Share (%)

Source Market	2000	2001	2002	2003	2004
Chinese mainland	28.7	32.4	41.4	54.5	56.1
Taiwan	18.1	17.6	14.7	11.9	9.5
Japan	10.5	9.7	8.5	5.6	9.5
USA	7.3	6.8	6.1	4.4	5.2
South Korea	2.8	3.1	2.8	2.4	4.8
Macau	3.4	3.9	3.2	2.9	2.5
Singapore	3.4	3.1	2.6	1.7	2.2
UK	2.8	2.6	2.3	1.8	2.1
Australia	2.7	2.4	2.1	1.6	1.9
Malaysia	2.4	2.1	1.9	1.3	1.9

Source： PATA 2006

2.4　Main source markets of tourism to Macau

Both Hong Kong and Taiwan have been among the traditional Source Markets of Macau tourism. Besides, there has been a rapid increase in Chinese mainland market share into Macau since China resumed sovereignty of Macau and the latter became a Special Administrative Region. Chinese mainland, Hong Kong and Taiwan are now the first 3 important source markets of Macau tourism, representing 95% of the total arrivals to Macau in 2004. The remaining arrivals mainly came from Japan, USA, Philippines, South Korea and UK (See Table 7).

Table 7   Macau's Main Source Markets and Their Market Share (%)

Source Market	2000	2001	2002	2003	2004
Chinese mainland	24.8	29.2	36.8	38.9	57.2
Hong Kong	54.1	50.6	44.2	48.3	30.3
Taiwan	14.3	14.1	13.3	8.6	7.7
Japan	1.6	1.4	1.2	0.7	0.7
USA	0.8	0.7	0.7	0.5	0.6
Philippines	0.5	0.5	0.5	0.4	0.5
South Korea	0.5	0.5	0.4	0.3	0.4
UK	0.5	0.4	0.4	0.3	0.3

Source: PATA 2006

With these observation and analyses, the 2 points that should be emphasized are as following:

• As far as the non-local Chinese tourist flows are concerned, Chinese mainland, Hong Kong, Macau and Taiwan to a great extent are and will continue to be important source markets with each other.

• With Chinese Mainland, Hong Kong, Macau and Taiwan as different destinations, their main foreign source markets are extremely similar.

3. Potentials of Chinese mainland, Hong Kong, Macau and Taiwan as source markets with each other

3.1   The potential of Chinese mainland as a source of outbound travelers to Hong Kong, Macau and Taiwan

The outbound tourism of China started in 1980s when residents of the mainland were allowed to travel to Hong Kong and Macau on purpose of visiting friends and relatives. Soon after that, with the development of the enlightened policy, China's outbound tourism entered the growth phase in 1990s. And the volume of outbound travelers has been increasing since then. As far as their destinations are concerned, Hong Kong and Macau have been the first 2 destinations among the top 10. One important reason for that is, with the handover of Hong Kong and Macau back to China, their relationship with the inland got much closer. And this has stimulated greatly the demand of inland residents to visit Hong Kong and Macau. [3]

According to PATA, during the 3 years of 2006, 2007 and 2008, the tourist arrivals to Hong Kong from Chinese mainland would increase by 5.33% annually, to 13868500 in 2006, 14680500 in 2007, and 15385280 in the year of 2008. The tourist arrivals to Macau from Chinese mainland would increase by average 10.25% annually, to 12165000 in 2006, 13593500 in 2007, and 14786800 in the year of 2008 (See Table 8). Because of the status quo of visitor exchange across the Taiwan straits and the attitude of the Taiwan authority toward the issue, it is hard at present to forecast the market volume of tourism to Taiwan from the mainland. Nevertheless, it is no doubt the mainland will quickly become the No. 1 source market of Taiwan tourism once the Taiwan authority gives up the regulation and allows residents of the mainland to visit Taiwan.

Table 8   The Potential Volume of the Mainland Travelers to Hong Kong, Macau and Taiwan 2006-2008

Destination	2004	2006	2007	2008	AAGR (%)
Hong Kong	12245862	13868500	14680500	15385280	5.33
Macau	9528739	12165000	13593500	14786800	10.25
Taiwan	na	na	na	na	na

Source: PATA 2006

3.2   The potential of Taiwan as a source of outbound travelers to Chinese mainland, Hong Kong, and Macau

According to PATA, during the 3 years of 2006, 2007 and 2008, the tourist arrivals to Mainland China from Taiwan would be increasing by average 7.77% annually, to 4.70 million in 2006, 5.09 million in 2007, and 5.43 million in the year of 2008. Within the same time span, the tourist arrivals to Hong Kong from Taiwan would increase by average 2.53% annually, to 2.20 million in 2006, 2.26 million in 2007, and 2.31 million in the year of 2008. And the tourist arrivals to Macau from Taiwan would increase by 15.13% annually, to 1.83 million in 2006, 2.13 million in 2007, and 2.43 million in 2008 (Table 9).

Table 9 The Potential Volume of Taiwan Travelers to Chinese mainland, Hong Kong and Macau 2006-2008

Destination	2004	2006	2007	2008	AAGR (%)
Chinese mainland	3685250	4697044	5093749	5454981	7.77
Hong Kong	2074795	2196094	2264513	2308539	2.53
Macau	1286949	1833820	2129040	2430820	15.13

Source: PATA 2006

3.3 The potential of Hong Kong as a source of outbound travelers to the mainland, Taiwan and Macau

PATA also forecasted that during the 3 years of 2006, 2007 and 2008, the tourist arrivals to the Mainland from Hong Kong would increase at an average annual growth rate of 3.45%, to 73.23 million in 2006, 75.74 million in 2007, and 78.37 million in the year of 2008. Those arrivals to Taiwan from Hong Kong would increase by average 6.16% annually, to 0.46million in 2006, 0.48 million in 2007, and 0.51 million in the year of 2008. And those arrivals to Macau from Hong Kong would increase at an average annual growth rate of 9.24% annually, to 6.13 million in 2006, 6.79 million in 2007, and 7.32 million in 2008 (Table 10).

Table 10 The Potential Volume of Hong Kong Travelers to the Mainland, Taiwan and Macau 2006-2008

Destination	2004	2006	2007	2008	AAGR (%)
Chinese mainland	66538862	73226000	75744000	78366000	3.45
Taiwan	417087	455384	483874	513195	6.16
Macau	5051059	6132600	6789800	7317700	9.24

Source: PATA 2006

3.4 The potential of Macau as a source of outbound travelers to the mainland, Hong Kong and Taiwan

With the methods suggested in Asia Pacific Tourism Forecasts 2006—2008 to convert the arrivals data for source markets to interpret the equivalent departures, it can be projected that during the 3 years of 2006, 2007 and 2008, the tourist arrivals to Hong Kong from Macau would increase by 6.75% annually, to 0.52 million in 2006, 0.56 million in 2007, and 0.59 million in the year of 2008. Yet it is difficult to predict the potential of Macau as a source of outbound travelers to the Mainland and Taiwan because of the data unavailable.

**Table 11 The Potential Volume of Macau Travelers to the Mainland,
Hong Kong and Taiwan 2006-2008**

Destination	2004	2006	2007	2008	AAGR (%)
Chinese mainland	na	na	na	na	na
Hong Kong	545266	516012	556119	588046	6.75
Taiwan	na	na	na	na	na

Source: Adapted from the data in *Asia Pacific Tourism Forecasts 2006-2008*

4. The way to develop tourism cooperation among the destinations across the Taiwan straits

With the above analyses, it seems clear that,

• The position of Asia Pacific region in world tourism has been rising.

• Northeast Asia plays an important part in the development of Asia Pacific tourism.

• The destination area with Chinese mainland, Hong Kong, Macau and Taiwan as a whole holds a dominant position in Northeast Asia tourism.

• Because of their close relations in geography, race and culture, Chinese mainland, Hong Kong, Macau and Taiwan as important source markets with each other have a huge potential of tourism development.

All these mean that it will be of great significance if these destinations can work together and strengthen their cooperation in developing tourism. Yet it will also be necessary to understand the possible problems in this regard. Firstly, there exist political constrains in visitor exchange between Taiwan and the mainland. Secondly, the administrative authorities of these destinations might have different positions, either in their motivations or in their efforts to support the development of tourism. All this means that it now seems too early to discuss the all-round cooperation of these destinations in developing tourism, particularly it is unrealistic to talk about their comprehensive tourism planning as one single destination.

With all these analyses, this author would like to argue that it seems more realistic at present that their cooperation should start with the inbound tourism marketing among the travel trades across the Taiwan straits with mutual benefits and all-win as principles.

It should be noticed that there is a co-petition among them as independent destinations. Because of the similarity in their foreign source markets, these destinations may play the role of either rival or ally. Rivals here mean those destinations whose success occurs at our expense. That is, the choice facing the tourist in the respective segment is to visit the rival destination or to visit us. Allies also covet the same market segment but, for various reasons, complement rather than diminish our efforts to attract these tourists. As pointed by J.R. Brent Ritchie, a well known scholar of tourism, '...two (or more) destinations may be allies because their geographical proximity enables packaging of the destinations in the same trip, or their collective marketing efforts may expand the size of the segment to the extent that both destinations experience more demand than they would if either of them was the only competitor ....' [4] Certainly, there may be other reasons for destinations to develop cooperation.

Like the goal of PATA in developing Asia Pacific tourism, the cooperation in the field of tourism marketing among destinations across the Taiwan straits should to some extent be established with the goal to attract more overseas visitors, existing and potential, from other parts of the world to come to visit these destinations, and to promote the local Chinese market traveling among these destinations. In this regard, there are at least two realistic strategies that seem worthy of consideration.

• To develop cooperative marketing targeted at their common source markets

The common source markets here mean in fact their foreign source markets, particularly those long haul markets from Europe and North America. The basic conditions for cooperative marketing in this regard are as the following. On one hand, because of their geographical proximity, homogeneous culture and complementary attractions, all of these enable packaging of these destinations in the same trip, especially as far as the long haul tourists from Europe and North America are concerned. On the other hand, in long haul tourism, tourists usually travel to more than one destination. As for the destinations across the Taiwan straits, the long haul tourists visiting any one of them may become a transfer market for other destinations within the area. Besides, many European and American travelers come to visit Asia because they are interested in oriental culture. Yet most of them do not know much about the Chinese culture, esp. its differences from other Asian cultures. If the destinations across the Taiwan straits can work together in market development with the authentic Chinese culture as the common theme, their collective marketing efforts may expand the size of the segment to the extent that all these destinations experience more demand.

• To develop cooperative marketing targeted at the local Chinese market across the Taiwan straits

This strategy also has a better foundation. As source markets with each other, residents in Chinese Mainland, Hong Kong, Macau and Taiwan are of the same race and share the same culture. This means there is no cultural barrier for them to travel within the area. And because their geographical proximity, the travel costs between these destinations are lower so that more local people can afford to travel within the area. In fact, the analyses and forecasts mentioned earlier have to a great extent suggested the necessity and importance for these destinations to develop cooperative marketing as source markets with each other. It is reasonable to believe that this kind of cooperation would be a cost effective way for the travel trades of these destinations to do marketing. And once the Taiwan authority gives up the regulation and allows the residents of the mainland to visit Taiwan, it will bring a new opportunity for the travel trades across the Taiwan straits to put this kind of cooperation into practice.

However, as far as the details in this respect, its organization and tasks in particular, are concerned, further exploration and research will be needed, especially by those who work in the travel trades across the Taiwan straits.

References

[1] Lindsay W Turner, Stephen F Witt. *Asia Pacific Tourism Forecasts 2006–2008* [M]. Beijing: China Tourism Press, 2006. 10–11, 127 (in Chinese, translated by Wang Xiangning).〔澳〕Lindsay W Turner〔英〕Stephen F Witt:《亚洲太平洋地区旅游业发展预测 2006~2008》[M], 王向宁主译, 北京: 中国旅游出版社, 2006 年 10 月 11 日, 第 127 页。

[2] Li Tianyuan. *Tourism(2nd edition)* [M]. Beijing: Higher Education Press, 2006. 252 李天元:《旅游学（第二版）》[M], 北京: 高等教育出版社, 2006, 第 252 页。

[3] Du Jiang, Dai Bin.Annual Report on Chinese Outbound Tourism 2004 [M].Beijing: Tourism Education Press, 2005.12–14. 杜江、戴斌:《中国出境旅游发展年度报告 2004》[M], 北京: 旅游教育出版社, 2005, 第 12~14 页。

[4] Ritchie B, Crouch,G. *The competitive destination* [M], Cambridge: CABI Publishing. 2003.27.

The Comparative Study on Friendly Degree of Internet Information between International Tourist City at Home and Abroad

Zhong Lina, Wu Bihu

(Peking University, Beijing 100871, China)

Abstract: With the development of tourism, many cities in China put forward the development goal of building an international tourist city. This paper tries to advance the friendly degree of international network marketing concept and analyze the differences in friendly degree of international network marketing between world renowned international tourist city and Chinese international tourist city by contrast the following areas: the total tourist internet information in English, internet English Tourist Information Quality, and the city's official tourist website. Though the comparison, this paper analyze the differences and similar between international tourist cities home and abroad when they attract global tourists online. This could help Chinese tourist cities build up an effective tourist information system through internet and attract more international tourists during the process of becoming a real international tourist city.

Key words: international tourism;tourist city website;tourist internet information

[About authors] Ms. Zhong Lina, Ph. D. candidate in the College of Urban and Environmental Sciences at Peking University, with research interests centered on issues related to the e-tourism and tourism planning. Mr.Wu Bihu, professor of tourism in the College of Urban and Environmental Sciences at Peking University, with research interests centered on issues related to the theoretical studies of tourism and tourism planning.

Summery

Tourist industry is an information-intensive Industry, information effective delivery and information circulation is its source of life-force. Information and guidance in Tourism Function System are not directly-produce economic benefit link, but they are actually the most important link to develop tourist market. Early in 1979, America scholar named Gunn put forward Tourism Function System conception which tells that Information and guidance is the key connected tourist with destination.

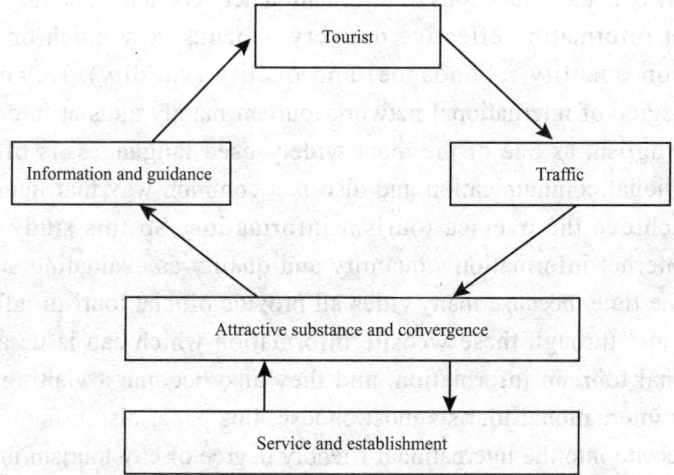

Figure 1   Tourism Function System (Gunn, 1979)

Especially in the international tourist industry which information is asymmetry, Information effective delivery and information guidance still should be a special concern of international tourism marketing. Appearance of internet brings about new opportunity in order to deliver international tourism information effectively. Through the internet browser, tourists can get the graphic-writing information about tourist destination, so this can accelerate tourist to make a tourism decision. At the same time, Internet becomes a strategic marketing place that numerous world tourist destinations announce tourism information and that lead to consumption. Many international tourist cities all have tourist information systems that processes of multi-language, quantity, types, plentiful function, tourist information system covering every aspect of information and internet tourist industry system to attract global tourist and potential tourist.

Along with the development of international tourist industry, many cities in China put forward the development goal of building an international tourist city, and make a quite in-depth discussion on tourist cities basic features and basic conditions of building international tourist city. However, international tourist cities marketing in china still keep traditional promoting ways. Comparing the foreign tourist cities, we have some distances in theory and practice of using internet to make oversea network marketing. This paper tries to bring forward the friendly degree conception of international network marketing and analyze whether tourism information which provided by different tourist cities network can actually play these functions of information delivery and marketing.

Internet information effective delivery depends very much on internet information quantity (abundance) and quality (validity). However, the friendly degree of international network tourism mainly aims at international tourists. English, as one of the most widely-used languages, is often used in international communication and also is a common way that international tourists achieve the oversea tourism information. So this study chooses English internet information's quantity and quality as evaluating standards. At the same time, because many cities all provide official tourism information websites, and through these website information which can issue abundant international tourism information, and they also become a visiting website that many international tourists must choose, this paper also bring every city official website into the international friendly degree of city tourism in order to make comparison, and we regard them as one of these important standards of measuring the international friendly degree of a city tourism information.

1. Case selection

1.1 Case city selection

To make sure the representative of this study, in the selection of the world tourism city, this paper makes reference of the ten leading of the world best tourist cities in 2006 selected though tourism investigation result of the global annul readers in American magazines-"travel and leisure" in every year, and this paper select five representative cities on region and language, such as: Florence, Bangkok, Sydney, Cape Town, New York, and select the first best tourist cities which are recommended for world in china, such as Beijing, Xian, shanghai and Guilin.

Table 1   World Best Tourist City in 2006

Place	Chinese Name	Country Name	Continent	Official Language
1	Florence	Italy	Europe	Italian
2	Rome	Italy (Capital)	Europe	Italian
3	Bangkok	Thailand(Capital)	Asia	Thai
4	Sydney	Australia(Capital)	Australia	English
5	Chiengmai	Thailand	Asia	Thai
6	Cape Town	South Africa (Legislation Capital)	Africa	South Africa Dutch, English
7	Buenos Aires	Argentina (Capital)	South America	Spanish, Italian, French, German
8	New York	America (Capital)	North America	English
9	Beirut	Lebanon	Asia City	Arabic
10	San Francisco	America	North America	English

The data source: http://www.travelandleisure.com/worldsbest/2006/results.cfm?cat=cities

1.2   Selection of search engine

Studying internet information material depends very much on selection of internet information search engine. According to American investigation report, it shows that 90% American use search engine to get information (yuan meng, 2005), 49.2% users among these all use Google as the preferred search engine, and Google is the top quality search engine in Argentina, Australia, Belgium, Brazil, Canada, Denmark, France, Germany, India, Mexico, Spain, Sweden, Switzerland British.Until 2005, the global users are up to 1 billion, and the global exclusive identity users of Google in every month are up to o.38 billion. So this study chose Google as representative of search engine about website information.

2. Comparative research of the friendly degree

2.1   English information gross quantity

English, as one of the most widely-used languages, and it is commonly-used language in international communication, using English to pass information can make the different language oversea tourists understand, so this paper regards gross quantity of English tourism information as an important index to measure the friendly degree of internet information system.

Using Google(www.google.com) search engine which is the maximal information coverage quantity in the world, users put some English Web Pages including Florence travel, Bangkok travel, Sydney travel, Cape Town travel, New York city travel (to distinguish New York state), Beijing travel, Xi'an travel, Shanghai travel, Guilin travel into advanced search to get on the whole English Web Pages quantity in these cities that provided by internet in the word. Although these information may has some definite shortage or is incompletely effective, we can make a rough comparison on information quality.

Table 2   Information Gross Quantity Comparison of English Tourism Website

rank	city	Key Words	Information Quantity	Country Name	Official Language
1	New York	New York City	29200000	America(Capital)	English
2	Sydney	Sydney	5790000	Australia (Capital)	English
3	Cape Town	Cape Town	3980000	Thailand(Capital)	Thai
4	Beijing	Beijing	3870000	China(Capital)	Chinese
5	Bangkok	Bangkok	2880000	Italy	Italian
6	Shanghai	Shanghai	2710000	China	Chinese
7	Florence	Florence	2300000	South Africa (Legislation Capital)	South Africa Dutch, English
8	Xi'an	Xi'an	315000	China	Chinese
9	Guilin	Guilin	93300	China	Chinese

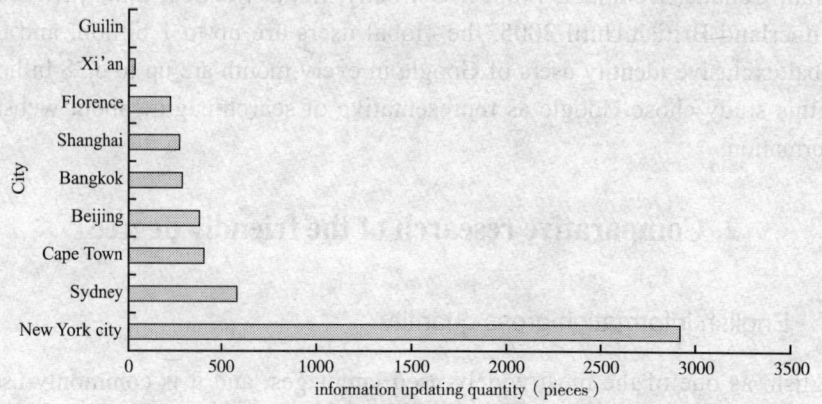

Figure 2   Comparative Chart of English Tourism Webpage Information Gross Quantity

From the comparison, we can easily find, among nine investigated cities, cities of ranking the top four are all national capitals that some English countries rank in front and that English tourism information quantity in Beijing is bottom of the rank.In the non-capital cities, Shanghai, Guilin and Xi'an separately rank 6, 8, and 9. Generally Speaking, the quantity of English tourism information is comparatively lesser, especially Guilin and Xi'an, the quantity of English tourism information is especially less than other cities because of itself less internationalization degree.

2.2   English information quality

In face of the large information in the web, the possible way of getting the oversea tourism information source for tourists is that uses the international search engine. According to an investigated result issued by prospect which is the marketing specialty service businessman of the earliest search engine in America in 2002, it is showed that, the users above 75% use search engine, 56.6% users only see the first 2 pages content of the search result, probably 16% users only see the former items of the search result, only 23% users examine the No.2 page content, the quantities of the users who see pages decline to 10.3%, the users who just see the first 3 pages only have 8.7% (Fen Jian Ying, 2002). These characters who use the search engine determine the tourism information quality of ranking in front of other results, and these qualities determine greatly that tourists make judgment of tourist cities English web information. So in the study of English information quality, it is a research object to filtrate first page information of choosing "city English name and travel" in Google as key words.

From the searching results, in the first page result, the tourism information in every international tourism city mainly roots in a few kinds of websites, as following: the city site in regional tourism information website; official website; city tourism website. This paper will make the special appraisement for the evaluation of official website of tourism city in a following part of this paper, this part make evaluations about information quality of a city website in regional tourism information website and information quality of special city tourist website.

2.2.1   The comparative research of information quality in city website of globality tourist information website

Tourism websites of ranking first and second sites named www.lonelyplanet.com and www.virtualtourist.com are the global tourism information portal websites to provide global tourism city English tourism information. As information platform, the information structure provided by them for every city is basically accordant. These information all reflect city information about tourist attraction, hotel, celebration, traffic and so on, however, from the information quality, every city have some differences. In this study, this paper choose information updating quantity as a main quality index to study the nearly tree months' updating quality difference of china and foreign countries international tourism cities in the two big global tourism information websites. It is showed that some cities, such as: New York, Bangkok and Sydney, their information quality and quantity in the global information website are better than other cities, this situation is also that global information websites make different orientation service for the difference attention of tourism destination. So establishing the internationalism tourism cities in china just depend on the information provided by internationalism tourism information website, this is very incomplete, these cities in China need to establish their tourism information website to boost good international reputation.

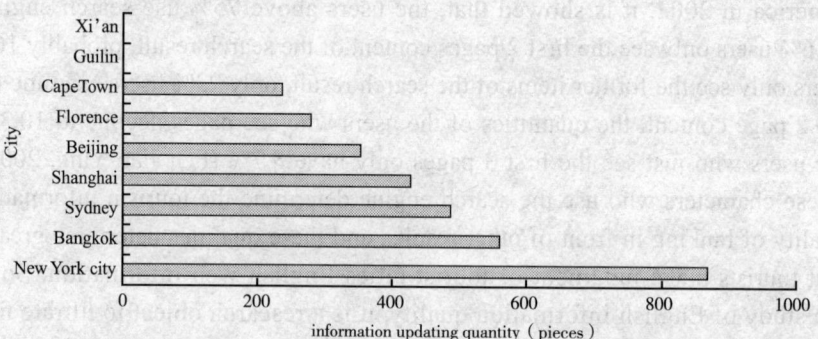

Figure 3 "www.lonelyplanet.com" Comparison Of Information Updating Quantity in International Tourist City Website

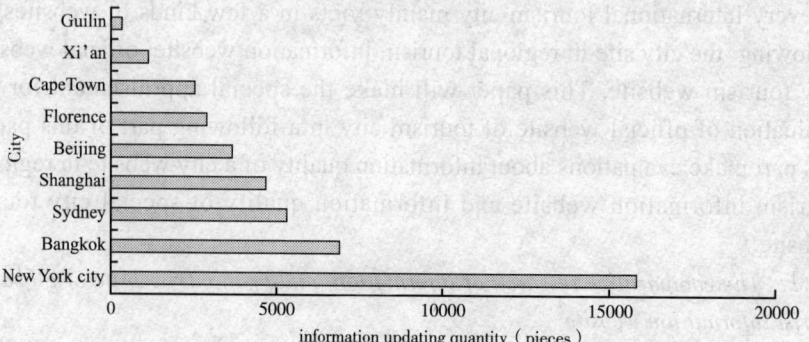

Figure 4 "www.virtualtourist.com" Comparison of Information Update Quantity in International Tourist City Website

2.2.2   Tourism information quality comparison in city English tourist website

Using Google to search city English tourism sites, according to the former 10 terms of statistic search results, these results show that 6 international tourism sites have their owns information site of city tourism, for example, Cape Town have five sites, Beijing have four sites, New York have two sites, Bangkok, Sydney and Guilin have separately one .This study choose Alexa rank as the index of website information quality evaluation, and after making rank analysis of searching websites with Alexa, it is found that rank and quantity in the city English tourism website have no correlation affiliation, for example 5 websites in Cape Town which rank in Alexa websites is relatively the bottom of the rank, and nearly these websites provided quoted tourist products and hotel tourist advanced booking by travel agency, but these effective information provided by travel agency is less, this situation is basically coherent with most of website in Beijing. The city tourism sites in New York Bangkok Beijing and Sydney are better in quality and have the front comprehensive rank in Alexa rank. This paper selects highly-ranked tourism websites in every city to make the following chart. Noteworthiness is that highly- ranked city tourism websites except Guilin and Cape Town, are tourism cities official tourism websites .It is obvious that tourists are comparatively admissive for official tourism website.

2.3   The comparison of the official English tourist website

According to six basic indexes that valued maturity of tourism destination website, they are: accessibility, experience character, validity, interactivity,

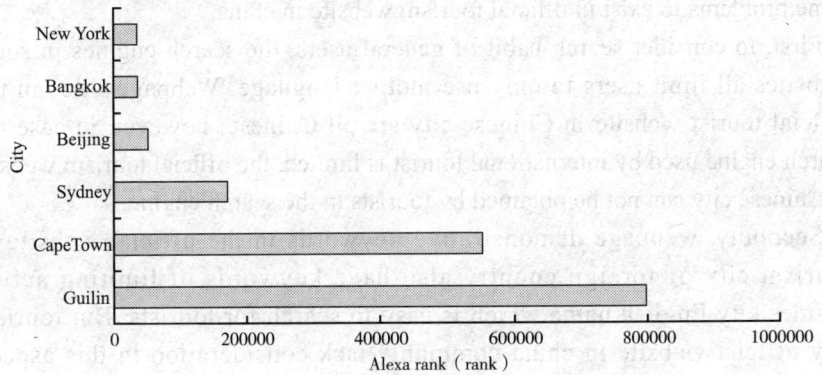

**Figure 5   Alexa Rank Comparison of Main English
Tourist Website in City**

business character, marketing ability, this paper use these indexes to make comparative study of nine international tourism cities official websites. (Zhong et al, 2007)

2.3.1 Accessibility

In this study, we use website rank of Google search engine as a main index to value its research ability. We put city English name into Google (www.google. com) to search all the web pages in order to calculate official web rank, it can see their official websites of the other 8 cities except shanghai are the top five, it is also right accessibility.

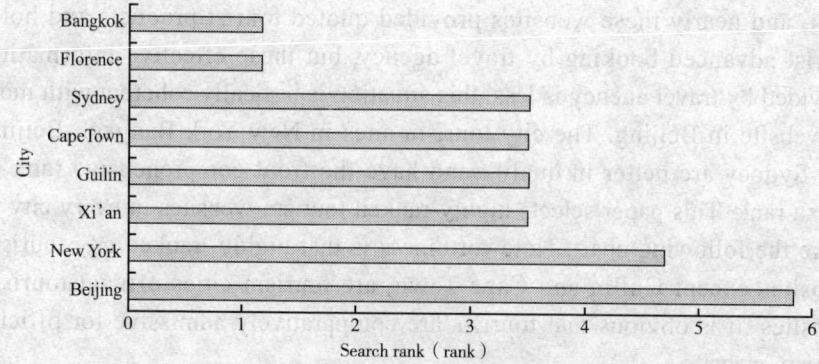

Figure 6　The Search Rank Comparison of Official English Tourist Website in International Tourist City

Although accessibility of tourism information official website in china and the website in foreign country are better, the deepen study at the base of this can find some problems to exist in official tourism website in china.

First, to consider search habit of general users, the search engines in some websites all limit users to only use mother language. Webpage titles in the official tourist website in Chinese city are all Chinese; however, in case the search engine used by international tourist is limited, the official tourism website in Chinese city can not be obtained by tourists in the search engine.

Secondly, webpage demonstrative keywords in the official website in tourism city of foreign country also have keywords of limiting action besides city English name which is easy to search for tourists. But tourism city official website in china commonly lack consideration in this aspect. The study find, using "city English name+travel" to search, the official tourism website of the other city in China also at least rank after 30 items except Beijing.

In addition, it is found that we cannot find the city official tourism website about Shanghai using "Shanghai" in English to search because the city official tourism website have not English edition.

2.3.2 Experience character

Functions of Website experience mainly use reasonable and considerable design (function, interface) to help net surfers conveniently and cozily to use the websites in order to get website information. This kind of website mainly represent these functions, such as multi-language version support, better content organizational structure, navigation system, interface style that is favorable to scan content and read, simple registration program and so on (Zhong et al, 2007). To speak from the content, content structure of international tourism city official English website is in term of travel, purchase, amusement, meal to organize. So, at the base of selecting multi-language edition and the different edition, this study regard browsing habit in the different country and the supporting degree of view in tasting beauty as major standard to value experience character of the official website of international tourist city, as following:

It is found that the official tourism website in New York and Florence separately have seven and six language editions, these websites have plenty consideration for language requirement of the international tourist, some websites in Xi'an and Guilin, at the base of Chinese-English edition, also consider their language requirement of the tourists who come from Japan and

Table 3 Multi-language Version Support Situation of The Official Tourist Website International Tourist City

	English	simplified Chinese	traditional Chinese	Japanese Edition	Korean Edition	Spanish	French	Italian	German
Shanghai	●								
Xi'an	●	●	●	●					
Guilin	●	●	●		●				
Beijing	●	●		●					
Florence	●		●			●	●	●	●
Bangkok	●			●					
Cape Town	●								
Sydney	●								
New York	●			●	●	●	●	●	●

"●" Represents that these websites provide these language edition.

Korea of major market areas, comparing these countries, the language editions provided by Beijing, Bangkok, Sydney and Cape Town are lesser.

This study also investigate that different language edition in official tourism website have consideration for user experience. It is found in this study that different language edition in the most of official website carry though change in the character of the content, but the information content and information exhibiting manner don't carry though change according to different user habit. Hereinto, only three city official tourism website have difference in the different language edition. The English edition of Xi'an website use red and yellow color which represent capital in china as basic color which differ quite vivid edition in Chinese. Chinese-English website in Beijing has difference in the interface structure in order to respect different Chinese-English reading habit, moreover, the official website in new York consider reading habit of German tourist, the German interface style have difference.

2.3.3 Validity

Website information character is determined by information quantity, information credibility degree and information update degree. The information credibility degree in the official tourism website is relatively higher, this study mainly make update information quantity of English website in the nearly 3 months as the main standard of validity. But in the study, we leave out Cape Town and shanghai, because the website in Cape Town is built in nearly 3 months, we can not get recently-update data, and at present that the English official website in Shanghai is still not built.

In the study, we found that the official tourism website in Beijing, Bangkok and New York keep the average updating quantity of one-two pieces of

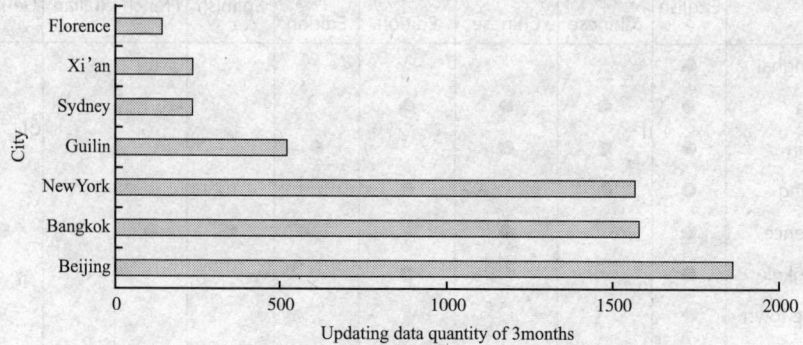

Figure 7   Updating Data Quantity of Official English Tourist Website in Every City in Recent Three Months

information in every day, the updating quantity in Guilin is up to 0.5 pieces in every day, but the updating quantities in Xi'an Sydney and Florence relatively are less.

2.3.4 Interactivity

Interactivity mainly reviews timely feedback information system of online tourist service and consulting that is provided by website. It is found in this study that international tourism city official website basically don't provide online timely -feedback service, and that they all resolve these communication problem among tourists using telephone, fax, E-mail and leave words system. The official website in New York and Beijing definitely give concrete addresses of underline tourist service center in order that tourist can get effective help in the entity address.

**Table 4   The Comparative Investigation of Interactivity Manner
in the Official English Tourism Website in Every City**

	Telephone	Message Board	Fax	E-mail	Forum
Xi'an	●				
Guilin	●				
Beijing	●		●	●	
Florence	●	●			
Bangkok	●	●	●	●	●
Cape Town	●	●	●	●	●
Sydney	●	●			●
New York	●		●	●	

"●" Represents these interactivity manners provided by these websites.

2.3.5 Business character

Business character mainly study website ability included providing real time and effective tourism product price information and supporting online electric business affairs. It is founded in the study that, all of international city official website provide online prearranged service, just have a difference in service scope, but hotel reservation is provided in all of cities, entertainment activity prearrangement is less. Official tourism in Sydney provided full-scale electric business activity from hotel, admission ticket of scenic spots, vehicle to tourist route, these can be prearranged, this website support freedom combination and pickings of route product, this website should be called city tourism electric system mirror.

Table 5   The Comparative Research of Business Character in the Official English Tourism Website in Every City

	Hotel Advance Booking	Route Advance Booking	Traffic Vehicle Advance Booking	Plane Ticket Advance Booking	Entertainment Activity Advance Booking
Xi'an	●				
Guilin	●				
Beijing			●		
Florence	●			●	
Bangkok	●	●		●	
Cape Town	●	●		●	
Sydney	●	●	●	●	●
New York	●	●		●	

" ● " Represents the business function provided by these websites.

2.3.6   Marketing ability

Website marketing ability mainly represents website external marketing which let more and more tourists visit them and remember them. In this study, we chose website pages with its friendly-link website as scale standard. After searching effective link of official English website in Google advanced search, we can get some research conclusions, as following: website link quantity in New York and Cape Town is comparatively large, its website external marketing ability and influence are accordingly higher.

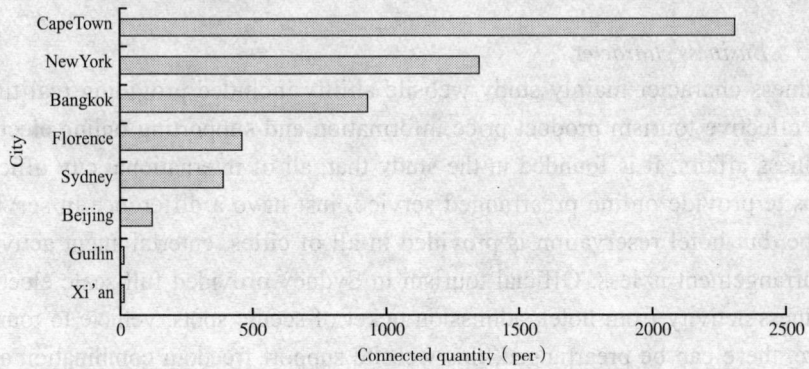

Figure 8   The Comparative Research of Marketing Ability in the Official English Tourism Website in Every City

3. Comparative research and appraisements

Making a conclusions of above-mentioned investigate result, we can get comparative research conclusions, as follow:

From information quality and quantity analysis, it is better than this function in china to express and transfer tourism information to international tourism via English website information in a whole.

From English information gross analysis, generally speaking, the higher the city internationalization is, the more its English information quantity is, this is closely linked with identity for English information and website utilization degree in city development. So, as such a city which internationalization degree is relatively low, but its international tourism development is comparatively outstanding, for example Sian, should still provide much more abundant English information in its city tourism destination information website to satisfy tourists need.

From English information quality analysis, it is not enough to provide China international tourism information by large-scale global tourism information platform. To convenience a large number of tourists in the world who are already used to use this sort of internet platform, these cities in china should actively enrich tourism information in global information platform. In the aspect of building city English tourism website, Access amount and web rank of foreign tourism city are higher than tourism city in China, it is reason that these experience of building city English tourism website in China, and that the aim of building is also much more utilitarian tendency, unitary information quality is not enough high.

In the six evaluated aspects of making comparison and analysis in English official tourism website, different city all have the different representations, among these, such as: Beijing in China, New York in America, Sydney in Australia, their comprehensive representations in every aspect are relatively excellence, and they should be good to use for reference for other cities, what this paper want to demonstrate is that, Shanghai, as an internationalization tourism city, have not itself official English website, it is a key problem that needed to be paid attention to in the process of building international tourism city in Shanghai.

References

[1] Feng Ying Jian. The understanding of the character of using search engine by users ［EB/
OL］. http://www.marketingman.net/wmtips/p126.htm. 2006–10–07. 冯英健.《了解用

户使用搜索引擎的特征》[EB/OL].

[2] Google company and the company information introduction. [EB/OL] .http://www.google. com/intl/zh−CN/corporate/index.html.2006−10−07.

[3] Yuan Meng. Web 2.0 the intellectual property problem of the era [EB/OL] .http://www. csip.org.cn/template/show_news.jsp?addr=upfiles/news/200512/20051201182254663233. htm. 2006−10−07. 袁萌.《Web 2.0 时代的知识产权问题》[EB/OL].

[4] CLARE G A. Tourism Planning [M] . NewYork:Crane Rusak, 1979: 58−73.

[5] World's best awards 2006 [EB/OL] .http://www.travelandleisure.com/worldsbest/2006/ results.cfm?cat=cities. 2006−10−07 .

[6] Zhong Lina, Shao J, Wu Bihu. Capability Maturity Model (CMM) to evaluate websites of Chinese 4A scenery spots [Z] . Honkong:The 3rd China Tourism Forum. 2006.

A Study on the Perceived Impacts of Historic & Cultural Ancient Town on Residents' Life Quality: A Case of Zhouzhuang in China

Guo Yingzhi, Pei Yanling, Ye Yunxia

(Department of Tourism, Fudan University, Shanghai 200433, China)

Abstract: First, this study took China historic & cultural ancient town Zhouzhuang as an example so as to study the perceived impacts of tourism development on local residents from the view of local residents of tourism destination. Second, the perceived evaluation of local residents' life quality after tourism development had been explored. Third, the relationship between evaluation of local residents' life quality and tourism perceived impacts was studied. Fourth, the support attitude of local residents for tourism development was indicated. The purpose of this study was to supply scientific base on local residents, active engagement in tourism development for local government.

Key words: resident; destination; life quality; evaluation; tourism impact; relationship

1. Introduction

More and more attention is being paid on the remarkable impacts of tourism

[About authors] Ms. Guo Yingzhi, Ph.D, associate professor of tourism in the Tourism Department at Fudan University, with research interests centered on issues related to the theoretical studies of tourism and marketing in travel and tourism. Ms. Pei Yanling and Ms. Ye Yunxia are master students in the Tourism Department at Fudan University.

development on local economy, employment, cultural exchanges and environmental improvement, especially on concrete interests gained by residents in matured tourist destinations during tourism development. Therefore, how to treat relationship between tourism destinations and tourist impact on the quality of life evaluation is a major concern while it is also playing a significant effect on developing a people-oriented society.

Along with the trend of economic globalization and China's full liberalization of foreign tourism in the past 20 years "China's Water Town" Zhouzhuang investing in the tourism industry are constantly increasing, the tourism Zhouzhuang have made rapid development on both quantity and quality. Located in the southwest of Kunshan City of Jiangsu Province in China, Zhouzhuang is adjacent to Shanghai in the east and to Suzhou in the west. It links with Shanghai-Nanjing Expressway in the north and Shanghai-Qingpu-Pingwang Express, and its rapid-flow port is an important hub on the waterway between Shanghai and Suzhou. Zhouzhuang is beloved with a long history and brilliant culture, which dates back to a relatively big water town in the first year of Yuanqing in Northern Song Dynasty (in the year of AD 1086). After 900 years of changes, its characteristics, "bridges, streams, cottages", which are unique of water town still sustain (Zhouzhuang Administration Government, 2007) [1].

Thanks to more financial investment into its tourism industry in recent two decades, Zhouzhuang has gained a lot of awards both at home and abroad, such as in China, "National Excellent Tourist Resorts of China", "Beautiful Counties with Fine Environment of China", "First National Historical and Cultural county", and "Highest National Environmental Prize", and from abroad, "UN Dubai Outstanding Example Awards in Residential Environment Improvement", "UNESCO Achievement Awards for World heritage preservation in Asian area", "Governmental awards in California, US" (Zhouzhuang Administration Government, 2006) [2]. Thanks to the development of tourism industry and promotion of its popularity among tourists, Zhouzhuang has become one of China's matured tourist destinations, with its economy, society, culture and environment greatly impacted by tourism industry. Residents here also have deep and all-round experience with development and changes of tourism industry, which provides great objective conditions for our empirical studies.

The significance of this research relies on two aspects. Theoretically, first, it is a guideline for dealing with the relationship between social development and improving people's living standards, persisting to the people-orientation principle, and actively solving most direct and practical, problems that are cared by people concerning their interests (Wen, 2007)[3]. Second, it has an academic value for establishing an all-round, harmonious and sustainable development theory, which would promote the overall development of economy, society and people. Third, it has a practical meaning in promoting employment opportunities and residents' living quality in tourist destinations (Wang, Wang, & Xing, 2005)[4]. Its findings can be used by the tourism administrative departments who rely on tourism to improve quality of life residents and raise the level of scientific basis for building a socialist harmonious society.

Practically, first, it will help residents in matured tourist destinations better understand the interaction between tourism development and their life quality. Second, it will assist tourism administrators identify the gains and losses of residents' interests, then realize the fine circulation among all interest parties in tourist destinations and enhance the development of tourism. Third, it will analyze the relationship between life evaluation of the tourist destination and the impact of tourism perception, fourth, it will help to study the supportive attitude towards the development of tourism and also to understand better about the interaction between their own quality of life and tourism, fifth, it will help tourism destination management departments to recognize the interests of residents aspirations and shortcomings and then realize the virtuous circle of promoting tourism development among the interests of the main tourist destinations. In fact, the ultimate goal of certification study is to mobilize initiatives of residents to participate in tourism destination development and then to provide a scientific basis for tourism administration departments initiate public policies which rely on tourism to improve the quality of life for residents.

2. Literature review

2.1 Domestic studies on impacts of tourist development on local residents' life quality

Firstly, many foreign scholars have made studies from the perspective of the residents on the perceived impact on tourism such as Long et. al. (1990) that affected residents Tourism Perception increased with tourism development and enhancing level[5]; McCool & Martin (1994) inspected the relationship between the perception of tourism impacts and residents depending on communities, the residents who are highly dependent on tourism development will be more awareness of the costs and impacts brought by tourism development [6]. Previous study shows that the perception of the residents for future community has the importance of supporting tourism development; Perdue, Long, & Allen (1990) recognized that those who get benefit from the tourism are more likely to support tourism follow-up development [7]. Based on social exchange theory, pointed that the wider theme of education and publicity campaign will help residents to get better understanding of destination tourism, and then the tourism industry will have greater support.

Secondly, domestic research in this area is also effective, such as Li, et. al. (1997) studied earlier on the perception of the residents of the domestic tourism impact. Huang [8], et. al. (2003) made a descriptive study on the early development of tourism destinations residents attitudes towards tourism [9]; Ying (2004) pointed that social reception comes mainly from the direct experience about the impact of tourism development of the local community residents [10]; Su & Lin (2004) used social surveys and market segmentation method making field surveys on the attitude and conduct of residents from towns of Xidi [11], Zhouzhuang and Jiuhuashan. Li & Cheng (2005) based on the tourists and local residents of Liuzhou City and used SPSS statistical software to analyze the findings of an evaluation of tourism awareness [12]. Chen, et. al. (2005) invested the perception and attitude of the Dunhuang Tourism urban residents towards the impact and development of tourism [13].

2.2 Overseas studies on the impacts of tourist development on local residents' life quality.

Firstly, Galbraith, a famous US New Institutional economist, first came up with the concept of life quality (Zhou, 2003) [14], which was then widely accepted and applied in various fields. The development of this concept has gone through the following stages by and large: threshold (the early 1900s to 1940s) (Rostow, 1971) [15], maturity (1950s to 1960s) (Galbraith, 1980) [16], and application (1960s to date) (Samuelson & Nordhaus, 2004) [17]. Upon entering 21st century, researches on life quality were mostly concentrated on its application in such fields as medicine (Schäfer, Riehle, Wichmann, & Ring, 2003 [18]; Fallowfield, 2005) [19], psychology (Castillón, Sendino, et. al, 2005) [20], sociology (Bradford, Rutherford, & John, 2004) and so forth, while relatively few were aimed at the impacts of tourism development on local residents' life quality [21].

Secondly, domestic scholars Li (1986) made a more objective evaluation on earlier the Western Economics "quality of life" study [22]; Chinese Academy of Social Science, "social development and social indicators" Taskforce that the quality of life in Chinese society is the development of an integrated system of indicators of an important component (Nan passengers, 1990) [23]; Bairi Rong and Zhang (2002) explored the connotation of quality of life, the importance of the evaluation system on the quality of life evaluation and the principle of choice on the quality of life, and bought forward that urbanization is the main reasons of affecting the lower the quality of life of our residents [24]; Zha Qifen and Song (2003) use analysis method on the quality of life of urban residents in Jiangsu Province for a comprehensive evaluation study [25]; Zhang (2005) made a comprehensive research on the quality of life issues from a different perspective [26].

2.3 Overseas and domestic studies on the impacts of tourist development on local residents' life quality mainly conclude two aspects

Firstly, researches on the impacts of tourist industry on residents' life. These studies include not only the positive and negative impacts (Carmichael, Peppard, & Boudreau, 1996 [27]; Perdue, Long, & Gustke, 1991) [28], but also economic (Davis, Allen, & Cosenza, 1988 [29]; Zhang, Wang, & Li 2003 [30]; Pan, Li, & Huang, 2003) [31], socio-cultural (Smith, 1989 [32]; Andereck & Vogt, 2000 [33]; Wang, 1999 [34]; Liu, 1998) [35], political (Kim & Crompton, 1990 [36]; Kim & Prideaux, 2003 [37]; Guo, Li, & Song, 2004 [38]; Qiu & Luo, 2003) [39] and environmental (Kuss & Grafe, 1985 [40]; Patterson, Gulden, Cousins, & Kraev, 2004 [41]; Peng, 1999) [42]; Liu, Lv, & Chen 2005) [43] impac ts of tourist industry on residents.

Secondly, researches on residents' perception and attitudes towards tourism development, including major categories of residents' perceptions (Evans, 1993 [44]; Weaver & Lawton, 2001 [45]; Lu, 1996 [46]; Huang & Wu, 2003) [47], the affected factors of residents' perceptions and attitudes (Andereck, Valentine, Knopf, & Vogt, 2005 [48]; Dyer, Gursoy, Sharma, & Carter, 2006 [49]; Zhan, Wang, Fan, 2005 [50]; Yan & Cheng, 2005) [51] and so on. The above research content indicates that few researches, both from home and abroad, directly discussed the impacts of tourism development on local residents' life quality, and further researches can benefit a lot from that in relevant fields.

The objective of this research is to probe into local residents' perceptions on the impacts of tourism development over their life quality in matured tourist destinations on the basis of literature review, taking Zhouzhuang, the first water-town in China as a case study, thus helping with tourism administration's public policies in enhancing residents' life quality through tourism development.

3. Data sources

Questionnaires were used to collect the data of residents' subjective perceptions on impacts of tourism development on their life quality; the design content was listed below. The first part is the survey about subjective effect of tourism on the life quality of destination residents; The second part is the survey about affected residents perception of the tourist destinations, the third part is the survey of evaluation residents on the perceived quality of life affected by the tourist destinations; The fourth part is the investigation of the overall judgment of residents toward quality of life affected by destination and the overall supportive attitude; the fifth part is the background information of the response, including demographic characteristics of survey residents and social attributes.

Firstly, the survey of subjective quality of life effect of destination residents by tourism, required that the respondents were asked on behalf of the family, income, health, entertainment, life, law and order and social impact of tourism by their own statements of degree; in order to get the accurate measurement of the quality of life residents by tourism, part of the problem reunification is "the development of tourism, the..." (for example, "tourism development, in your opinion, attracts more investment to the local tourism industry"; "Tourism development, makes your day-to-day contact with their families more closely than before", etc.), then let respondents make judgment according to their own perception, and thus gained observational data about impact of tourism destination on subjective quality of life of residents, finally conducted factor analysis.

Secondly, observational data about the perceptual impact of residents by tourism development, is based on a questionnaire survey of tourism destinations residents perception survey, respondents were asked questions about the development of the tourism economy, social and cultural tourism, tourism environment and tourism impact of 29 political issues, part of the questions were asked like "the development of tourism, the..."(for example, "tourism development, in your opinion, attracts more investment to the local tourism industry"; "Tourism development, makes your day-to-day contact with their families more closely than before", etc.), then let respondents make judgment according to their own perception, and thus gained observational data about impact of tourism destination on subjective quality of life of residents, finally conducted factor analysis.

Respondents were required to rate their level of agreement with each on a 7-point Likert scale from strongly disagree (1) to strongly agree (7). In this way, it will be clearer to weigh effected perception of tourism development.

According to contacts level between residents and tourists destinations and exchanges and economic dependence on tourism destinations, residents were divided into four levels: the first tier is people who have recurrent, sustained contact with tourists, their economic income is mainly depended on tourism, including hotels and restaurants, travel agencies, tourism company employees in individual industrial and commercial households; The second tier is the people who have irregular contact with tourists, their economic income is partly depended on tourism, including business, entertainment and transportation industry employees; The third tier is the people who have regular, sustained contact with tourists, their economic income is less dependent on tourists residents, referring to ordinary residents and villagers; the fourth tier is the people who have less contact with tourists, the economy is not dependent on tourism income residents, including retirees, students and teachers, and other non-employees in the tourism sector. As each layer of residents has no definite data as a reference, the four parts are sampled in accordance with the quota-sampling.

The survey period was the first two weeks in March of 2006. 320 questionnaires were spread out then, among which 307 were usable. The survey was conducted by questionnaires and face-to-face interviews. In the first period of investigation, residents that economically mainly or partly depend on tourism industry were investigated, such as employees in hotels, travel agencies, and transportation departments, and self-employed laborers in tourism industry. In the second period, those common residents or villagers, who frequently or continuously have connections with tourists, but economically less dependent on tourism industry, were investigated in randomly selected three villages located in sightseeing and commercial areas in central town. In the third period, employees in non-tourist departments or industries, such as governmental officials, students, teachers, etc were investigated. Among all the people investigated, 46.7% were male, 63.8% were younger than 30 years old, 82.1% were from families with a monthly income lower than 3999 RMB, 49.5% had received higher education than college, 39.7% were from families composed of more than 4 people, 62.2% were from families with 3 generations, 57.4% had a residence of more than 20 years, 68% were urban citizens, and, 47.3% were involved in tourism-related professions (Table 1).

Table 1   The Demographic Profiles of Respondent Residents in Zhouzhuang (N=307)

Variable	Category	% of respondents	Variable	Category	% of respondents
gender	male	46.7%	Living place	City and town	68.0%
	female	53.3%		Village and farm	32.0%
age	18 or below	15.0%	Family revenue per month	Less than 1999 RMB	38.8%
	19-30	48.8%		2000-3999 RMB	43.3%
	31-50	30.7%		4000-5999 RMB	6.2%
	51 or above	5.5%		6000RMB or above	11.7%
Educationallevel	Primary or below	2.0%	Family members	1-2 persons	6.5%
	Secondary schools	48.5%		3 persons	53.7%
	undergraduate	49.2		4-5 persons	36.1%
	postgraduate	0.3		6 persons or above	3.6%
livingtime	Five years or below	10.7%	occupation	Hotels\travel services\tourism company employees in individual industrial and commercial households	23.6%
	5-10years	12.0%			
	11-20years	20.0%		business, entertainment and transportation industry employees	23.7%
	21-30years	22.7%		ordinary residents or villagers	14.0%
	31years or above	34.7		retirees, students and teachers	38.7%
generations	2 generations	30.9%			
	3 generations	62.2%			
	4 generations or above	6.9%			

4. Study results

4.1 Residents' perceptions of the impacts of tourism on their life quality in a matured tourist destination

Table 2 shows the mean scores and standard deviations of the variables, which represented residents' perceptions of their life quality after the developing of tourism. From this table, respondents' perceptions and evaluations on the items of living quality can be analyzed.

Among the 33 selection criteria in Table 2, items related to residents' family lives were viewed moderate with mean scores greater than 4.00, and the item of overall satisfaction with family lives was rated relatively high, with a mean score of 4.33. By contrast, items related to income were viewed low, especially the item of "After the developing of tourism, you have less spending on life necessities (e.g. foods and clothing)", with the mean score of 2.32, representing a lower identification with a decreasing Engel Coefficient due to tourism development, in other words, residents still regarded Engel Coefficient there high. The mean score of the item "After the developing of tourism, your Income (income) has increased" was 3.76, lower than the average mean, with a standard deviation of 2.02, which indicates that respondents had quite different opinions on their income and also a variation in benefits among different groups in the destination. Items related to health were viewed as moderate, while exceptionally, the item "After the developing of tourism, you have better medical conditions and services in local hospitals" was rated high with the mean score of 4.30, which shows that the development of tourism had brought with the promotion of local infrastructure and services, particularly that of medical conditions. Most respondents had high recognition of items related to recreation, especially of the item "After the developing of tourism, you have more choices of entertainment", with the mean score of 4.50, which indicates that tourism industry served not only tourists but also local residents. Respondents viewed highest toward items related to promotion in life facilities and services, all with mean scores greater than 4.50, for instance, generally they believed that "After the developing of tourism, life facilities (e.g. post offices, water supply, electricity supply) are improved and have more vegetable and food supply in local area". Residents possess a strong sense of identification and belonging is concerned, and items of "After the developing of tourism, you have a stronger sense of pride as a local resident than before" and "After the developing of

Table 2   Statistics of Observational Variables of Interviewed Residents' Subjective Life Quality

Variables	Mean	Standard Deviation	Rank
Have more contacts with family members	4.02	1.74	18
Have more harmonious relationship with family members	4.07	1.68	16
Satisfied with family life	4.33	1.67	11
Income increased	3.76	2.02	22
Less spending on life necessities (e.g. foods and clothing)	2.32	1.62	32
More spending on entertainment and recreation	4.56	1.71	6
More spending on education	4.40	1.73	10
Price of real estate reduced	2.01	1.60	33
Better health conditions than before	3.73	1.65	23
Better medical conditions and services in local hospitals	4.30	1.73	12
Get more in-time and effective medical treatment than before	4.08	1.72	15
Have better moods	4.02	1.67	18
Have more choices of entertainment	4.50	1.67	8
Have more relaxation and fun in life	4.27	1.64	13
Have more time to travel	3.47	1.65	26
Have more time to meet friends	3.71	1.61	24
Have more time for enjoying radio and TV programs	3.83	1.84	21
Life facilities (e.g. post offices, water supply, electricity supply) are improved	4.96	1.68	1
More vegetable and food supply in local area	4.63	1.72	4
Family appliances (air conditioners, computers, etc) were improved	4.60	1.78	5
More comfortableness of living in this tourist area	4.05	1.73	17
I can see a brighter future living in this tourist area	4.71	1.73	3
Stronger sense of pride as a local resident	4.93	1.60	2
Local security services are improved	3.44	1.86	27
Less chance of seeing local policemen	3.05	1.73	31
Less crimes in local area	3.06	1.90	30
Personal safety and property safety are better secured	3.33	1.71	28
Tourists wouldn't worry about security problems	3.29	1.73	29
Neighborhood became better	3.97	1.57	20
More time together with local friends and relatives	3.68	1.68	25
People are more active in communication	4.27	1.66	13
Have wider social network than before	4.56	1.76	6
Better social relationship than before	4.41	1.65	9

tourism, you can see a brighter future living in this tourist area" had mean scores
of 4.93, 4.71, respectively. The items related to public security were perceived
lower, for example, mean scores of items of "After the developing of tourism,
you have less crimes in local area" and "After the developing of tourism, you
have less chance of seeing local policemen" were 3.06, 3.05, which indicates
that as the development of tourism, with the rush of tourists, pressures and
problems on local security came as well. Items of social contacts were mainly
perceived high, three of which "After the developing of tourism, you have wider
social network than before", "After the developing of tourism, neighborhood
became better", "After the developing of tourism, People are more active in
communication" had mean scores of 4.56, 4.41, 4.27 respectively.

The 33 observational variables in Table 2 reflected residents' perceptions and
evaluations on items of life quality from various aspects, however, they were
too complicated because of possible relationships between some variables. For
construct a general evaluation model of residents' evaluation of life quality,
factor analysis was adopted to eliminate subordinate elements while combine
significant information. In order to ensure the suitability of using factor analysis,
Bartlett Test of Sphericity and Kaiser Meyer Olkin Test (KMO) were employed.

**Table 3   Results of Bartlett Test of Sphericity and Kaiser Meyer
Olkin Test on Life Quality Items**

Tests for suitability of using factor analysis		Test Score
Kaiser Meyer Olkin Test		.882
Bartlett Test of Sphericity	Approx. Chi-Square	4695.275
	df	528
	Sig.	.000

In Table 3, the KMO measure of sampling adequacy was found to be 0.882.
At the same time, the Bartlett's test of sphericity was 4695.275 (at the degree
of freedom 528) with 0.000 significance level lower than that set before, which
showed that zero hypothesis could be rejected and marked linear correlations
between interactive matrixes in the sample. These figures suggested that the use
of factor analysis was appropriate for this study.

Principal component analysis was applied to extract original factors. With
varimax rotation, eigenvalue with score above 1 were selected, whose name
was given in accordance with that with a relatively higher factor loading. Table
4 shows the eigenvalue, percentage of explained variance and total explained

Table 4 Results of Factor Analysis of Residents' Perceptions of Life Quality

Factor	Eigenvalue	Variable	Factor Loading	Explained Variance(%)	Total Explained Variance(%)	Reliability Coefficient α
1. Family Life	3.296	Have more harmonious relationship with family members	0.743	9.988	9.988	0.812
		Have more contacts with family members	0.724			
		Satisfied with family life	0.690			
		Income increased	0.673			
2. Life Pattern	3.022	Have more time to travel	0.694	9.158	19.146	0.792
		Better health conditions than before	0.645			
		Have more time to meet friends	0.612			
		More time together with local friends and relatives	0.521			
		Have more relaxation and fun in life	0.506			
3. Social Communication	2.980	People are more active in communication	0.825	9.032	28.178	0.852
		Have wider social network than before	0.777			
		Better social relationship than before	0.754			
		More time together with local friends and relatives	0.514			
4. Community Identification	2.959	I can see a brighter future living in this tourist area	0.732	8.966	37.144	0.763
		Stronger sense of pride as a local resident	0.699			
		More comfortableness of living in this tourist area	0.635			

Continued Table 4

Factor	Eigenvalue	Variable	Factor Loading	Explained Variance(%)	Total Explained Variance(%)	Reliability Coefficient α
5. Public Security	2.922	Tourists wouldn't worry about security problems	0.775	8.853	45.997	0.820
		Local security services are improved	0.717			
		Personal safety and property safety are better secured	0.713			
		Less crimes in local area	0.616			
		More spending on education	0.570			
6. Life Facility	2.665	Life facilities (e.g. post offices, water supply, electricity supply) are improved	0.565	8.076	54.073	0.698
		Family appliances(air conditioners, computers, etc) were improved	0.560			
		Have more time for enjoying radio and TV programs	0.547			
		More vegetable and food supply in local area	0.534			
		Less chance of seeing local policemen	0.660			
7. Consumption Structure	2.015	Price of real estate reduced	0.657	6.106	60.180	0.618
		Less spending on life necessities (e.g. foods and clothing)	0.562			
8. Recreation Consumption	1.491	More spending on entertainment and recreation	0.831	4.517	64.697	0.523
		Have more choices of entertainment	0.536			

Notes: 1. Selecting ways:Principal Components Analysis.

2. Circumvolving ways: Varimax with Kaiser Normalization.

variance of each factor. Besides, factors selected in Tables 4-9 were those with factor loading greater than 0.50, and generally speaking, items with absolute factor loadings greater than 0.3 were significant. 33 variables were combined into 8 main factors, each of which was an eigenvalue greater than 1. These 8 main factors represented 65.7% of the total explained variance. In other words, these 8 factors comprising 33 variables covered 64.7% information of original variables. Reliability analysis was then conducted to test the internal reliability of these 33 items. Cronbach's alpha was calculated as 0.9285, which indicated than the scale constituted by these 33 variables was acceptable and reliable.

Results of factor analysis were as follows. Factor 1 had relatively higher loadings on income, relationship within the family and satisfaction with family life, reflecting the impact level of tourism on residents' family life and could be named as Family Life Factor. Factor 2 was also an integrated one with relatively higher loading on recreation and health, reflecting the impact level of tourism on residents' life patterns and could be named as Life Pattern Factor. Factor 3 had relatively higher loadings on local residents' social communication frequencies and scopes, reflecting the impact level of tourism on residents' social communication and could be named as Social Communication Factor. Factor 4 had relatively higher loadings on community dependence, community identification and sense of belonging and could be named Community Identification Factor. Factor 5 had relatively higher loadings on the increasing security pressures because of tourism development and could be named as Public Security Factor. Factor 6 mainly showed that tourism development brought about positive effects on local infrastructure, public services, and food supplies and so on and could be named as Life Facility Factor. Factor 7 mainly reflected residents' consumption structure variations and the impact of tourism development on Engel Coefficient, which could be named as Consumption Structure Factor. Factor 8 reflected the impacts of tourism development on residents' choices and expense in recreation and could be named as Recreation Consumption Factor.

4.2 Destination residents' perceptions of the impacts of tourism development

In addition to investigating the impact of tourists' perception on the residents of the quality of life, the paper also surveyed destination residents' perceptions of the impacts of tourism development. Table 5 identifies the mean scores and standard deviations of the variables, which represented residents' perceptions of the tourism development. From this table, respondents' perceptions and

**Table 5   Statistics of observational variables of interviewed
residents' tourism perceptions**

Variables	Mean	Standard deviation	Rank
Need more working opportunities	4.75	1.61	15
Get more working opportunities from tourism	4.85	1.64	13
Attract more investments for tourists	5.36	1.49	6
Expense of tourists increased	5.49	1.43	5
Price of product and service increased	5.62	1.49	1
More investments by government for tourism development	5.61	1.37	2
More revenue brought to the government by tourism industry	5.56	1.43	3
Increased family revenue promoted by tourism industry	3.88	1.88	25
Positive effects brought by tourism industry to local cultural	4.90	1.66	12
Positive effects brought by tourists life style	4.22	1.58	22
It's worthy to use your tax to improve public facilities	4.61	1.70	17
Local social circumstances improved	3.55	1.62	28
If you remove from the town now, you will find it hard to leave	4.35	1.97	19
Your awareness of protecting local cultural enhanced	5.17	1.54	9
Your awareness of protecting architectural monuments	5.26	1.52	7
Vandalism acts to the property of destination residents reduced	4.33	1.70	20
Destination residents could participate more community activities	4.93	1.60	10
Cultural qualities of destination residents improved	4.29	1.59	21
After tourism development, you are more likely to understand their cultural through making friends with them	4.91	1.73	11
Better local transportation	4.50	1.96	18
Local pollution (garbage, noise, etc.) reduced	2.96	1.84	29
Better protection for local tourism resources	4.12	1.71	23
Huge using space for development	3.82	1.95	26
Stong confidence for local economical environment	3.95	1.91	24
Have more curiosity to get news and events	4.78	1.61	14
Higher level government offer more support	4.62	1.45	16
Overseas Chinese' visiting to China strengthen their feelings	5.24	1.53	8
Open a window to know Chinese for foreigners	5.51	1.45	4
Better protection for privacy	3.77	1.54	27

evaluations on the items can be analyzed.

According toTable5, interviewed residents had stronger perception for the economic impact of tourism, especially after tourism development, "local goods and services prices increased" (5.62), "the government invest a lot of funds for the development of tourism" (5.61), "tourism industry brings more income to the government compared with other local industries" (5.56), "tourist spending in the local place increased" (5.49), "tourism industry attract more investment"(5.36), and on this aspect, interviewed residents of Zhouzhuang on the development of the tourism industry agree that the economic benefits, commonly hold the perception that the tourism development with the economic benefits, but it also indicated that "with the development of tourism, your household income increased," the recognition is 3.88 This shows that there is income gap of the interviewed residents between expectation and reality. Interviewed residents for the social impacts of tourism perception tends middle, but most of the residents agree with the "tourism bring a positive impact to the local cultural" (4.90) and more recognition "it's worthy to improve public facilities with their own taxes" (4.61). For the cultural impact of tourism, the perception of residents is strong, they generally reflected "enhance consciousness of the protection of architectural monuments" (5.26), "strengthened awareness of protecting local cultural" (5.17), and that "destination residents can participate in more cultural activities" (4.93), to some extent, the development of tourism has brought to local interviewed residents the attention to the value of tourism resources, so that the development of tourism also has the educational effect for residents. Interviewed residents had strong negative perception on tourism impact on environment, they generally do not agree that "with the development of tourism, local pollution (garbage, noise, etc.) reduced" (2.96), nor do they believe that "there is a great local land space" (3.82), they are pessimistic about the prospects on the protection of the ecological environment. The political impact of the tourism, the perception of interviewed residents is higher, they strongly agree with the "Open a window to know Chinese for foreigners" (5.51), and agree that "overseas Chinese or Chinese visiting to China strengthen their feelings" (5.24), but for a large number of inbound tourists, it's concerned about the personal life will be interfered to a certain extent.

Similarly, in order to ensure the suitability of using factor analysis, Bartlett Test of Sphericity and Kaiser Meyer Olkin Test (KMO) were employed.

As shown in Table 6, the sample value of 0.832 KMO, greater than 0.8, that is more suitable for factor analysis. In addition Bartlett ball test statistics for 3290.585 (DOF for 406), significantly less than the set of standards, the

**Table 6   Results of Bartlett Test of Sphericity and Kaiser Meyer Olkin
Test on Life Perception Items**

Tests for suitability of using factor analysis		Test Score
KMO Test		.832
Bartlett Test of Sphericity	Approx. Chi-Square	3290.585
	df	406
	Sig.	.000

original hypothesis can be rejected, on behalf of the parent group there is an obvious correlation matrix of the linear relationship, so that it's suitable for factor analysis. Through the above test, the sample data is in accordance with the observation of the precondition for factor analysis, the analysis can be done further study.

Principal component analysis was applied to extract original factors. With varimax rotation, eigenvalue with score above 1 were selected, whose name was given in accordance with that with a relatively higher factor loading. Table 7 shows the eigenvalue, percentage of explained variance and total explained variance of each factor. Besides, factors selected those with factor loading greater than 0.50, and generally speaking, items with absolute factor loadings greater than 0.3 were significant (Guo, 1999). [52] 29 variables were combined into 8 main factors, each of which was an eigenvalue greater than 1. These 8 main factors represented 62.7% of the total explained variance. In other words, these 8 factors comprising 29 variables covered 62.7% information of original variables. Reliability analysis was then conducted to test the internal reliability of these 29 items. Cronbach's alpha was calculated as 0.8926, which indicated than the scale constituted by these 29 variables was acceptable and reliable.

Results of factor analysis were as follows. Factor 1 had relatively higher loadings on cultural protection awareness, architecture protection awareness and participation for cultural activities, reflecting the impact level of tourism on residents' cultural protection and could be named as cultural perception factor. Factor 2 was also an integrated one with relatively higher loading on local circumstance improvement, community dependence and increased revenue, reflecting the impact level of tourism on community dependence and could be named as community perception factor. Factor 3 had relatively higher loadings on local residents' environmental effect, reflecting the impact level of tourism on environment and could be named as environment perception Factor. Factor 4 had relatively higher loadings on increased government tourism revenue and

Table 7 Factor Loadings and Explained Variance of Main Factors of residents' Perceptions of tourism development

Factor	Eigenvalue	Variable	Factor Loading	Explained Variance (%)	Total Explained Variance (%)	Reliability Coefficient α
1. Cultural perception	3.619	The awareness of protecting local cultural enhanced	0.820	12.479	12.479	0.795
		The awareness of protection architectural monuments enhanced	0.754			
		Vandalism acts to the property of destination residents reduced	0.693			
		Destination residents could participate more community activities	0.612			
2. Community perception	2.520	Local social circumstances improved	0.714	8.690	21.169	0.670
		If you remove from the town now, you will find it hard to leave	0.658			
		Increased family revenue by tourism	0.637			
		It's worthy to use your own tax to improve public facilities	0.564			
3. Environment perception	2.431	Better protection for tourism resources	0.763	8.384	29.553	0.681
		Local pollution reduced	0.748			
		More confidence for the prospect of eco-environment	0.611			
		More space for land using	0.540			

Continued Table 7

Factor	Eigenvalue	Variable	Factor Loading	Explained Variance (%)	Total Explained Variance (%)	Reliability Coefficient α
4. Benefit perception	2.340	A lot of funds invested by government	0.739	8.069	37.622	0.668
		Tourism brought more revenue to government compared to other industries	0.726			
		Price of product and service increased	0.585			
5. Employment perception	2.275	More working opportunities brought by tourism industry	0.741	7.846	45.469	0.697
		More investments brought by tourism development	0.696			
		Expenses of tourists increased	0.606			
		Want more working opportunities	0.542			
6. Policy perception	1.966	Higher level government offer support to local government	0.748	6.778	52.247	0.670
		More initiatives to know news and events	0.735			
7. Price perception	1.545	Price of local product and service increased	0.577	5.328	57.575	0.340
		Cultural quality of destination residents enhanced	0.519			
8. Privacy perception	1.497	Better privacy protection	0.820	5.161	62.736	—

Notes: 1. Selecting ways: Principal Components Analysis.

2. Circumvolving ways: Varimax with Kaiser Normalization

price and could be named benefit perception factor. Factor 5 had relatively higher loadings on working opportunities brought by tourism and could be named as employment perception factor. Factor 6 mainly showed that tourism development brought about positive effects on politic participation and public services and could be named as policy perception factor. Factor 7 mainly reflected the perception of residents about local price, which could be named as price perception factor. Factor 8 reflected the impacts of tourism development on residents' privacy and could be named as privacy perception factor.

4.3   The correlation analysis of perception of tourism effect and life quality evaluation of destination residents

The focus of this study is whether there is a relationship between tourism impact perception and life quality of residents. According to the previous analysis, the eight factors can explain the quality of life of residents in the perceptual evaluation which contained 64.7% information of the original variables, but can also explain the tourism impact perception variables which contained 62.7% information. In other ways, this 16 principal components factor values can represent evaluation of life quality and tourism impact perception. Table 8 horizontal listed eight main perception component factors related to the tourism impact of the residents, and longitudinal listed the eight main evaluation component factors related the quality life evaluation of residents. Using Pearson analysis, while the coefficient is Pearson correlation coefficient, significant probability (Sig. (2—tailed)) expressed bilateral significant test probability can be expressed in Table 8 and Figure 1.

Research findings can be summarized from the analysis of Table 8 and Figure 1.

Firstly, there is a significant positive correlation among evaluation of life quality and community perception of tourism impact, price impact of tourism and privacy perception. This shows when interviewed residents evaluate life quality, they consider about the effect on community, price and privacy of tourism.

Secondly, there is a significant positive correlation among evaluation of life quality and community dependence, self belonging feeling、 enhanced environment protection awareness and higher level government policy, this shows the development of tourism enhance the recognized and proud awareness of living place of destination residents and makes them be treasured about local resources and environment, at the same time, they expect more policy support from government and in this case, there will be more casual life style and

Table 8 The correlation analysis of perception of tourism effect and life quality evaluation of destination residents

Items		Cultural perception	Community perception	Environment perception	Benefit perception	Employment perception	Policy perception	Price perception	Privacy perception
1. Family life	Pearson coefficient	.069	.224(**)	.021	.023	-.107	.060	.215(**)	.306(**)
	Significant probability	.247	.000	.728	.700	.074	.317	.000	.000
2. Life style	Pearson coefficient	-.008	.184(**)	.158(**)	-.072	-.118(*)	.198(**)	-.073	.081
	Significant probability	.900	.002	.008	.229	.048	.001	.220	.174
3. Social communication	Pearson coefficient	.179(**)	.256(**)	-.032	.074	.146(*)	.177(**)	-.062	.123(*)
	Significant probability	.003	.000	.596	.215	.014	.003	.300	.040
4. Community identification	Pearson coefficient	.253(**)	.286(**)	.118(*)	.197(**)	.312(**)	.163(**)	.026	-.048
	Significant probability	.000	.000	.048	.001	.000	.006	.665	.424
5. Public security	Pearson coefficient	.046	.213(**)	.247(**)	-.127(*)	.089	.164(**)	-.071	.039
	Significant probability	.442	.000	.000	.033	.135	.006	.237	.513
6. Life facility	Pearson coefficient	.108	.044	.002	.184(**)	.044	.017	.152(*)	.208(**)
	Significant probability	.070	.460	.974	.002	.458	.775	.010	.000
7. Consumption structure	Pearson coefficient	-.250(**)	.222(**)	.084	-.195(**)	-.027	-.012	-.063	.171(**)
	Significant probability	.000	.000	.161	.001	.653	.835	.293	.004
8. Recreation consumption	Pearson coefficient	.091	.026	.071	-.041	.096	-.033	.148(*)	-.029
	Significant probability	.128	.664	.234	.490	.109	.582	.013	.623

Notes: 1. Use Pearson correlation analysis

2. ** correlation is significant at 0.01 level (Bilateral test)

3. * correlation is significant at 0.05 level (Bilateral test)

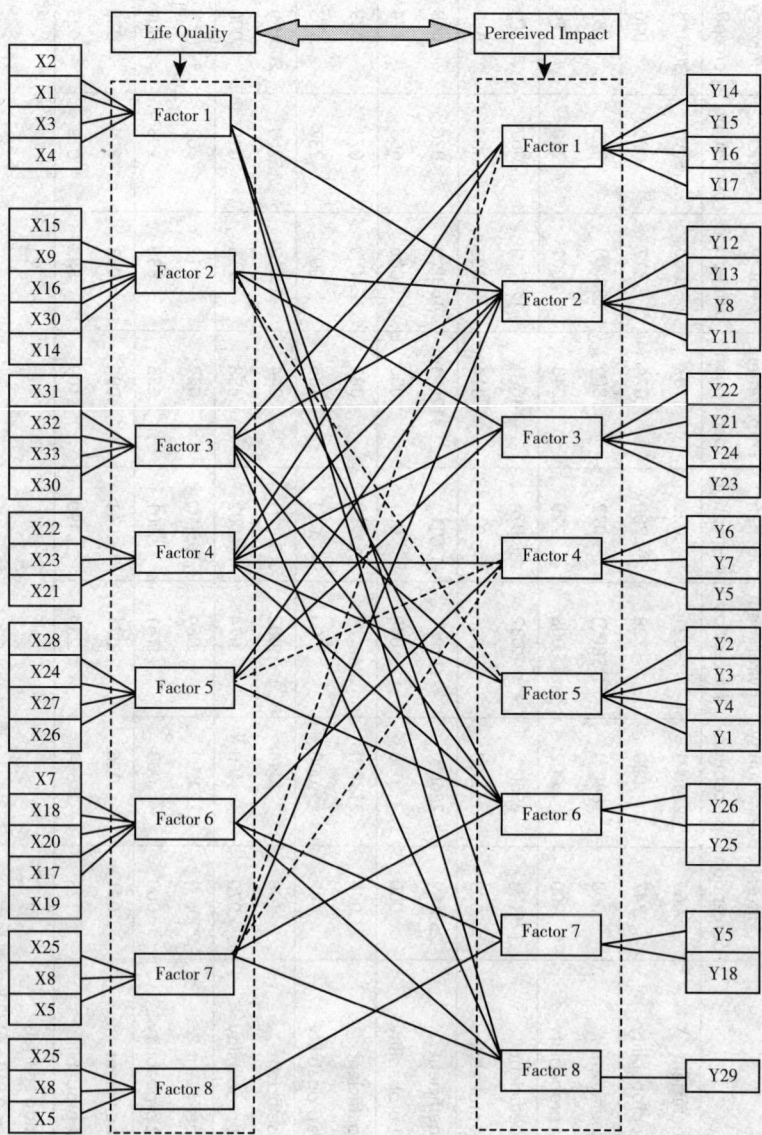

Figure 1   The correlation analysis of perception of tourism effect and life quality evaluation of destination residents

Notes: 1. Solid line means a positive correlation.

2. Dotted line means a negative correlation

3. Thick line means correlation is significant of 0.01 level

4. Thread line means correlation is significant of the 0.05 level

entertainment time and activities.

Thirdly, there is a significant positive correlation among evaluation of social communication and cultural effect of tourism、enhanced community dependence feeling and higher level government policy support, this shows enhanced cultural awareness of destination residents、dependence feeling for living place of destination residents and policy support from government brought by the development of tourism, this makes them have more confidence for future harmony life.

Fourthly, there is a significant positive correlation among evaluation of community recognition, cultural effect of tourism、enhances community dependence feeling, increase family revenue, more working opportunities and higher level government policy support, this shows again that enhance self-belong and proud feeling for community brought by the development of tourism, with increased family revenue and government support policies, they will hold stronger recognition and dependence for living place and future life.

Fifthly, there is a significant positive correlation among evaluation of community security and community dependence feeling and environment effect brought by environment, this shows residents evaluate community security based on the perception of environment changes, the more they are dependent on local community and environment, the more they are concerned about security.

Sixthly, there is a significant positive correlation among evaluation of life facilities and benefit changes brought by tourism, this shows that residents agree increased family revenue brought by the development of tourism, they also agree that life facilities get improvement and life necessaries get sufficient supply. In this way, only increased consuming ability can improve the gratification of product and service of tourists.

Seventhly, there is a significant positive correlation among consumption structure and privacy changes and community dependence feeling, this shows that the more residents are dependent on living places, the more they'd like their personal life be not disturbed, they also expect price of estate reduced and there are more free revenue for consumption except food and accommodation.

Eighthly, there is a significant positive correlation among evaluation of entertainment consumption and price effect of the development of tourism, this shows that residents realize the price increased with tourists swarm into the destination, they also hold the view that the consumption of entertainment increased compared before.

At the same time, there is a significant negative correlation between the evaluation of consumption structure of residents and cultural effect brought by tourism, this shows on one hand, the awareness of protecting local cultural enhanced by tourism and there are more opportunities to participate cultural activities, on the other hand, the price of entertainment resources and activities is too high, they expect the price to reduce. At the same time, there is also a significant negative correlation between the evaluation of consumption structure and increased economic benefic brought by tourism, this may show government get great amount of revenue so that it should share with destination residents, in detail they should reduce the price of estate and other price line of public facilities for residents.

Then we may make a test for two variables to prove the above consequences, and the value of two variables can be got through questionnaires. Ask interviewed residents "the development of tourism generally make a positive effect on the local economy and cultural" "the development of tourism generally make a positive effect on life quality of residents", the interviewed residents give the mark according to their own attitude, the formal question is to test the general perception of tourism impact on residents, the later one is to get the general evaluation of life quality after the development of tourism. The correlation of two is listed below in Table 9. Analyze the significant correlation of two issues could prove the above analysis result.

Table 9   Correlation analysis of the general tourism impact perception and evaluation of life quality of interviewed residents

		General perception of tourism impact	General evaluation of life quality
General perception of tourism impact	Pearson coefficient	1	.471(**)
	Bilateral test related probability	.	.000
General evaluation of life quality	Pearson coefficient	.471(**)	1
	Bilateral test related probability	.000	.

Note: ** means correlation is significant of 0.01

4.4   Correlated factors on support factors of destination residents towards tourism

Use correlation analysis to test the relationship between the support factors of destination residents towards tourism and perception of tourism impact and evaluation of life quality. In Figure 2 there is a positive correlation among tourism impact perception、life quality evaluation and the level of support to

tourism, especially to the residents who hold strong perception level of tourism
positive impact and life quality evaluation (both index have the mean value
at 4.00), this kind of positive correlation is more obvious. At the same time,
if the level of support to tourism of local residents is high (the mean value is
above 4.00), then the level of tourism positive impact perception is high, and
the evaluation of life quality is high, the above two have small distinction, this
could be reflected in the Figure 2, after the mean value 4.00, two curves have
very small distinction; if the level of support to tourism of local residents is not
high (the mean value is below 4.00), then the level of tourism positive impact
perception is not high, and the evaluation life quality is not high too, this could
also be reflected in the table, before the mean value 4.00, two curves have very
distinctive differences.

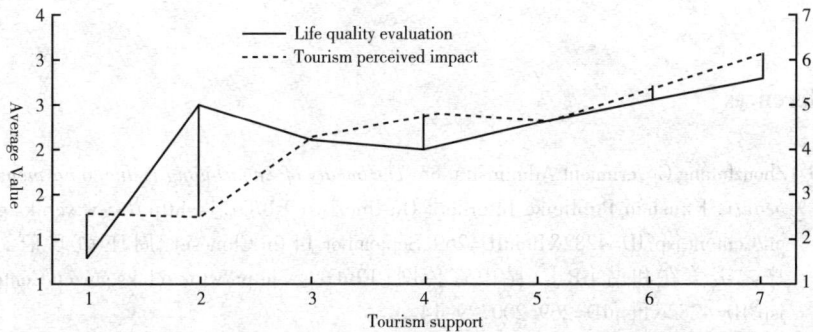

**Figure 2   Life quality evaluation、tourism impact perception
and tourism supporting relationship**

5. Conclusion and limitation

In conclusion, paying attention to perception evaluation to life quality of
tourism destination tourists, is also the essence of widely participation to
economical and social life for residents, the aim is to improve the results of
economic development, enhance the perception evaluation to life quality of
tourism destination tourists, enforce the control and execute ability of tourism
government in solving various possible problems, and as a result realize richness
and equality of destination residents and their families. Therefore, in constituting
a socialist harmony society, one of the key points is to realize the enhancement
and improvement about the perception evaluation of life quality of tourism
destination residents, currently, some governments have realize the importance

of this issue, and they have started to solve and improve this problem through developing the destination. In near future, the development of tourism may greatly promote and improve the perception evaluation to life quality of tourism destination residents.

With the rapid tourism development in Zhouzhuang, quality of local residents' life quality would further improve. There would be stronger demand for tourism development and better life quality for local residents, and would emerge more new features pending to explore. Therefore, further research on local residents' life quality improved by tourism development in Zhouzhuang should collect updated data and investigate further information as updated samples and investigation information advancing with the time would be more helpful to study the continuously expanding and changing residents' life quality and tourism perceived impacts.

References

[1] Zhouzhuang Government Administration. *The bueaty of Zhouzhuang is due to harmonious beauty*. Kunshan Publicize Internet［On line］.［EB/OL］. http://www.xcb.ks.gov.cn/Content.jsp?ID=4282&ItemID=269, September 14 (in Chinese). 周庄镇政府. 周庄之美美在和谐［R］. 昆山宣传网［EB/OL］. http://www.xcb.ks.gov.cn/Content.jsp?ID=4282&ItemID=269, 2007-9-14.

[2] Zhouzhuang Government Administration. *Summary Report of Zhouzhuang Town in 2006*, December 31 (in Chinese). 周庄镇政府. 周庄镇 2006 年工作总结［R］. 2006-12-31.

[3] Wen J. Government Report on the Fifth Meeting of the Tenth National People's Congress of China［On Line］. 2007-03-05. *www.sohu.com* (in Chinese). 温家宝:《十届全国人大五次会议政府工作报告》［R］, 2007-03-05. www.sohu.com

[4] Wang Zixin, Wang Yucheng, Xing Huibin. Tourism Impact Research Process［J］. *Tourism Magazine*, 2005, 2: 90–95 (in Chinese). 王子新、王玉成、邢慧斌:《旅游影响研究进展》［J］,《旅游学刊》2005 年 2 月，第 90~95 页。

[5] Long PT, Perdue R, Allen L. Rural resident tourism perceptions and attitudes by community level of tourism［J］. *Journal of Travel Research*, 1990, 28(3): 3–9.

[6] McCool SF, Martin, S. R. Community attachment and attitudes toward tourism development［J］. *Journal of Travel Research*, 1994, 32(3),29–34.

[7] Perdue RR, Long PT, Allen L. Resident support for tourism development［J］. *Annals of Tourism Research*, 1990, 17(4): 586–599.

[8] Li Yougen, Zhao Xiping, Zou Huiping Residents perception of tourism impact［J］. *Psychology Process*, 1997, 2: 21–27 (in Chinese). 李有根、赵西萍、邹慧萍:《居民对旅游影响的知觉》［J］,《心理学动态》1997 年第 2 期，第 21~27 页。

[9] Huang Jie, Wu Zanke. A Study of attitude by local residents of Tourist Destinations Towards Tourism Impact [J]. *Tourism Tribune*, 2003, 18(6): 84–89 (in Chinese). 黄洁、吴赞科:《目的地居民对旅游影响的认知态度研究——以浙江省兰溪市诸葛、长乐村为例》[J],《走向世界》, 2003 年第 6 期, 84–89.

[10] Ying Tianyu. The application of social representation theory in tourism research [J]. *Tourism Magazine*, 2004, 1: 87–92 (in Chinese). 应天煜:《浅议社会表象理论 (Social Representation Theory) 在旅游学研究中的应用》[J], 2004 年 1 月《旅游学刊》, 第 87~92 页。

[11] Su Qin, Lin Bingyao. Classification of the residents of China's tourism based on the attitude and behavior–Xidi, Zhouzhuang and Jiuhua Mountain [J]. *Geological Research*, 2004, 23(1): 104–114 (in Chinese). 苏勤、林炳耀:《基于态度与行为的我国旅游地居民的类型划分——以西递、周庄、九华山为例》[J],《地理研究》, 2004 年第 1 期, 104–114.

[12] Yan Lihong, Cheng Daopin. Analysis of industrial tourism perception social–based in Liuzhou city [J]. *Journal of Guilin Institute of Technology*, 2005, 3: 398–401 (in Chinese). 颜丽虹、程道品:《基于 SPSS 的工业旅游感知分析》[J],《桂林工学院学报》2005 年 3 月, 第 398~401 页。

[13] Chen Yongsheng, Wang Nai'ang, Fan Juanjuan. The study on residents' perceptions and attitude of Dunhuang city [J]. *Journal of Historical Geography*, 2005, 2: 66–71 (in Chinese). 谌永生、王乃昂、范娟娟等:《敦煌市居民旅游感知及态度研究》[J],《人文地理》, 2005 年第 2 期, 第 66~71 页。

[14] Zhou Changcheng. *Life quality in China: current status and evaluation* [M]. Beijing: Social Science Literature Press, 2003, 6 (in Chinese). 周长城:《中国生活质量:现状与评价》[M], 北京:社会科学文献出版社, 2003, 6.

[15] Rostow WW. *Politics and the Stages of Growth* [M]. Cambridge: Cambridge University Press, 1971, 13.

[16] Galbraith JK. *Wealthy Society* [M]. Beijing: Shangwu Press, 1980. 55–60. (in Chinese). J. K. Galbraith. 富裕社会 [M]. 北京:商务印书馆, 1980, 第 55~60 页。

[17] PA Samuelson, William D. Nordhaus. *Economics* (17th Edition) [M]. Beijing: Posts & Telecommunications Press, 2004, 301 (in Chinese). PA Samuelson, William D. Nordhaus. 经济学 (第 17 版) [M]. 北京:人民邮件出版社, 2004, 第 301 页。

[18] Torsten Schäfer, Angelina Riehle, H. Erich Wichmann, Johannes Ring. Alternative medicine and allergies: Life satisfaction, health locus of control and quality of life [J]. *Journal of Psychosomatic Research*, 2003, 55(6): 543–546.

[19] Fallowfield L. Acceptance of adjuvant therapy and quality of life issues [J]. *The Breast*, 2005, 14(6): 612–616.

[20] Pilar Guallar–Castillón, Áurea Redondo Sendino. Differences in Quality of Life between Women and Men in the Older Population of Spain [J]. *Social Science & Medicine*, 2005, 60(6): 1229–1240.

[21] Roger Bradford, Donna L Rutherford, Alexandra John. Quality of life in Young people: ratings and factor structure of the quality of life profile–Adolescent version [J]. *Journal of Adolescence*, 2002, 25(3): 261–274.

[22] Li Yining. *Socialist political economics* [M]. Beijing: Commercial Press, 1986, 523–525 (in Chinese). 厉以宁：《社会主义政治经济学》[M]，北京：商务印书馆，1986，第 523~525 页。

[23] Ye Nanke. The new connotation of social development: domestic and foreign "quality of life" study [J]. *Social Science Review*, 1990, 4: 35–38 (in Chinese). 叶南客：《社会发展的新内涵：国内外"生活质量"研究简述》[J]，社会科学述评，1990 年第 4 期，第 35~38 页。

[24] Bai Rirong., Zhang Hao A number of issues about assessment for quality of life [J]. *Journal of the CPC Shanxi Provincial Committee Party School*, 2002, 1: 45–46 (in Chinese). 白日荣、张浩：《生活质量评价若干问题探讨》[J]，《中共山西省委党校学报》2002 年 1 月，第 45~46 页。

[25] Zha Qifen, Xu Wensong. Comprehensive evaluation studies on the quality of life of urban residents in Jiangsu Province [J]. *Journal of Jiangsu University* (Social Sciences Edition), 2003, 4: 106–109 (in Chinese). 查奇芬、徐文松：《江苏省城市居民生活质量的综合评价研究》[J]，2003 年 4 月《江苏大学学报》(社会科学版)，第 106–109 页。

[26] Zhang Leisheng. Life quality: the scientific concept of development inherent requirements [J]. *Teaching and Research*, 2005, 2: 19–23 (in Chinese). 张雷声《生活质量：科学发展观的内在要求》[J]，《教学与研究》，2005 年第 2 期，第 19~23 页。

[27] Carmichael BA, Peppard DM, Boudreau FA. Mega–resort on my doorstep: local resident attitudes toward Foxwoods Casino and casino gambling on nearby Indian reservation land [J]. *Journal of Travel Research*, 1996, 34(3), 9–16.

[28] Perdue RR, Long PT, Gustke LD. The effects of tourism development on objective indicators of local quality of life [A]. In: Travel and Tourism Research Association. Tourism: Building Credibility for a Credible Industry [C]. *Proceedings of the 22nd Annual TTRA International Conference*. Salt Lake City, 1991, 191–201.

[29] Davis D, J Allen, R Cosenza. Segmenting local residents by their attitudes, interests and opinions towards tourism [J]. *Journal of Travel Research*, 1988, 27(3): 2–8.

[30] Zhang Fan, Wang Leizhen, Li Chunguang. Research on the contribution level of tourism industry to society and economics in Qinhuangdao city [A]. In: *China Tourism Development during 2002–2004: Analysis and Expectations* [C]. Beijing: Social Science Literature Press, 2003, 24–27. (in Chinese). 张帆、王雷震、李春光等：《旅游业对秦皇岛市社会经济的贡献度研究》[A]，见《2002~2004 年中国旅游发展：分析与预测》[C]. 北京：社会科学文献出版社，2003，第 24~27 页。

[31] Pan Jianmin, Li Zzhaorong, Huang Jin. Research on contributions of tourism on national economy in Guangxi [A]. In: China Tourism Development during 2002–2004: Analysis and Expectations [C]. Beijing: *Social Science Literature Press*, 2003, 28–31. (in Chinese). 潘建民、李肇荣、黄进：《旅游业对广西国民经济的贡献度研究》[A]. 《2002~2004 年中国旅游发展：分析与预测》[C]. 北京：社会科学文献出版社，2003，第 28~31 页。

[32] Smith VL. Hosts and Guests: *The anthropology of tourism* (2nd edition) [M]. Philadelphia: University of Pennsylvania Press, 1989.

[33] Andereck K, C Vogt. The relationship between residents' attitudes toward tourism and tourism development options [J] . *Journal of Travel Research*, 2000, 39(1): 27–36.

[34] Wang Xuehua. The tourism impacts on social culture [J] . *Journal of Guilin Institute of Tourism*, 1999, 2: 60–63. (in Chinese). 王雪华:《论旅游的社会文化影响》[J],《桂林旅游高等专科学校学报》1999 年 2 月, 第 60~63 页。

[35] Liu Zhaoping. Further analysis of social and culture impacts of tourism on destinations: tracking survey of tourism development in Yesanpo [J] . *Tourism Tribune*, 1998, 1: 50–54 (in Chinese). 刘赵平:《再论旅游对接待地的社会文化影响》[J],《旅游学刊》1998 年 1 月, 第 50~54 页。

[36] Kim Y, Crompton JL.. Role of tourism in uifying the two Koreas [J] . *Annals of Tourism Research*, 1990, 7(3): 353–366.

[37] Samuel Seongseop Kim, Bruce Prideaux. Tourism, peace, politics and ideology: Impacts of the Mt. Gumgang tour project in the Korean Peninsula [J] . *Tourism Management*, 2003, 24(6): 675–685.

[38] Guo Yingzhi, Li Lei, Song Shuling. Current status and cooperation strategies of tourism market between Mainland China and Taiwan [J] . *Tourism Research and Practice*, 2004, 4: 46–49 (in Chinese). 郭英之、李雷、宋书玲:《海峡两岸旅游市场发展现状与合作对策》[J],《旅游研究与实践》, 2004 年第 4 期, 第 46~49 页。

[39] Qiu Xiaohui, Luo Yufeng. Analysis of impacts of hosting Olympic on tourism [J] . *The Journal of Sport History and Culture*, 2003, 7: 26–27 (in Chinese). 邱小慧、骆玉峰:《主办奥运会对旅游的影响分析》[J],《体育文化导刊》2003 年 7 月, 第 26~27 页。

[40] Kuss RF, Grafe AR. Effects of recreation trampling on natural area vegetation [J] . *Journal of Leisure Research*, 1985, 17(3): 165–183.

[41] Trista Patterson, Tim Gulden, Ken Cousins, Egor Kraev. Integrating environmental, social and economic systems: A dynamic model of tourism in Dominica [J] . *Ecological Modelling*, 2004, 175(2): 121–136.

[42] Peng Xing. Current situation, reason and countermeasures the environmental problem on Wulingyuan scenic spot. *Journal of Changde Teachers University*, 1999, 4, 178–185 (in Chinese). 彭翔:《武陵源风景名胜区环境问题的现状、成因及对策》[J],《常德师范学院学报》(社科版), 1999 年 4 月, 第 178~185 页。

[43] Liu Yang, Lv Yihe, Chen Lixiang. The impact assessment of ecotourism in nature reserves: progress and implications [J] . *Journal of Natural Resources*, 2005, 5, 771–779 (in Chinese). 刘洋、吕一河、陈利顶等:《自然保护区生态旅游影响评价:进展与启示》[J],《自然资源学报》2005 年 5 月, 第 771~779 页。

[44] Evans TR. Residents perceptions of tourism in New Zealand [D] . *Masters of Commerce Thesis*, University of Otago, Dunedin, 1993.

[45] Weaver BD, Lawton JL. Resident perceptions in the urban–rural fringe [J] . *Annals of Tourism Research*, 2001, 28(2), 439–458.

[46] Lu Lin. Studies on attitudes of residents in tourist destinations: taking south Anhui tourism area as an example [J] . *Journal of Natural Resources*, 1996, 11(4), 377–382 (in Chinese). 陆林:《旅游地居民态度调查研究》[J],《自然资源学报》1996 年 11 月第 4 期, 第 377~382 页。

[47] Huang Jie. Wu Zanke. A Study of attitude by local residents of Tourist Destinations Towards Tourism Impact[J]. *Tourism Tribune*, 2003, 18(6), 84–89 (in Chinese). 黄洁、吴赞科:《目的地居民对旅游影响的认知态度研究——以浙江省兰溪市诸葛、长乐村为例》[J],《走向世界》，2003 年第 6 期：第 84~89 页。

[48] Kathleen L. Andereck, Karin M. Valentine, Richard C. Knopf, & Christine A. Vogt. Residents' perceptions of community tourism impacts[J]. *Annals of Tourism Research*, 2005, 32(4), 1056–1076.

[49] Pam Dyer, Dogan Gursoy, Bishnu Sharma, & Jennifer Carter. Structural modeling of resident perceptions of tourism and associated development on the sunshine coast, Australia[J]. *Tourism Management*, 2006, 28(2), 409–422

[50] Zhan Yongsheng, Wang Nai'ang, Fan Juanjuan. A study on residents' perceptions and attitudes toward tourism in the Dunhuang city[J]. *Human Geography*, 2005, 2, 66–71 (in Chinese). 谌永生、王乃昂、范娟娟等:《敦煌市居民旅游感知及态度研究》[J],《人文地理》，2005 年第 2 期，第 66~71 页。

[51] Yan Lihong, Cheng Daopin. Analysis of industrial tourism perception social–based in Liuzhou city[J]. *Journal of Guilin Institute of Technology*, 2005, 3, 398–401 (in Chinese). 颜丽虹、程道品:《基于 SPSS 的工业旅游感知分析》[J],《桂林工学院学报》2005 年 3 月，第 398~401 页。

[52] Guo Zhigang. *Social Statistics Methodology: Application of SPSS*[M]. Beijing: China, 1999. (in Chinese). 郭志刚:《社会统计分析方法——SPSS 软件应用》[M],北京：中国人民大学出版社，1999 年，第 99~101 页。

Study on the Spatial Structure and Cause of Leisure Tourism Peculiar Street in Shanghai

Song Changhai[1], Lou Jiajun[2]

(1. Department of Economic Management, STIEI, Shanghai 201411, China;
2. Department of Tourism, ECNU, Shanghai 200062, China)

Abstract: Leisure tourism peculiar street, an important role in developing urban tourism industry of Shanghai, is increasingly drawn attention and taken into great account nowadays. Based on positive research of the structure of those streets and an explicit knowledge of its current spatial layout as well as its structure characteristics, this article aims to interpret every possible cause for its multiplicity, which then may provide a scientific basis for its development in depth and reasonable adjustment.

Key words: leisure tourism peculiar street; spatial structure; cause; Shanghai

1. Introduction

Leisure tourism peculiar street is an innovative form of developing urban tourism resources, and also an important carrier of urban tourism development. There are

[About authors] Mr. Song Changhai(1981-), master, assistant of Economic Management of Shanghai Technical Institute of Electronics & Information, with research interests centred on urban tourism, leisure and city exhibition. E-mail: yjssch@sina.com; Mr. Lou Jiajun(1957-), Ph.D., dean and professor of Department of Tourism of East China Normal University, with research interests centered on urban tourism, city leisure and management of tourism company.

many prestigious leisure tourism peculiar streets reputed in the tourism markets at home and abroad, such as Wang Fu Jin and Tianqiao Bridge in Beijing, Xin Jie Kou and Confucius Temple in Nanjing, Guanqian Street in Suzhou, and Nanjing Road and Yuyuan Garden in Shanghai, etc. With the rapid development of urban tourism, various leisure tourism peculiar streets showing urban style and taste have received more and more attention of management and operating units. From horizontal comparison among domestic cities, the development of leisure tourism peculiar streets in Shanghai are particularly prominent.

2. Researches on peculiar streets and the development of Shanghai peculiar streets

The building and the development of leisure tourism peculiar streets is a new topic in the development of modern urban tourism. At present, the study on leisure tourism peculiar streets (hereafter referred as peculiar streets) has acquired a lot of progress. Internationally, the research on urban peculiar streets mainly concentrated in such fields as urban architecture, urban planning, urban tourism, and so on. For instance, Einar Lillebye (1996) discussed the internal relations between the buildings and formation of blocks from a historical perspective[1]; Assenna Todorova, Shoichiro Asakawa and Tetsuya Aikoh (2004) studied on the reasonable configuration of the infrastructure of urban blocks[2]; Stephen W Litvin (2005) mainly discussed relationship between rebuilding old urban areas and tourism development[3]. In the domestic researches, most studies focus on commercial pedestrian streets and tourism peculiar streets. In the research of commercial peculiar streets, Tang Fanming (2005) concerns the status and role of commercial pedestrian streets in the tourism industry[4]; Zheng Xiaoshan (2004) explored the design of commercial pedestrian streets[5]; Jiao Sheng (2005) analyses the development model of commercial streets from the rebuilding of old urban areas[6]; Tang Yue (2004) mainly discussed the relationship of the development of commercial streets and tourism[7]. In addition, Bao Jigang etc discussed pedestrian streets as a special form of urban RBD[8]; speaking from the researches on tourism peculiar streets, many scholars chose Shanghai as the research subject, such as Yao Kunyi (2000)[9] and Chen Jikun (2001)[10] described the functional position of Shanghai Urban leisure tourism peculiar streets; Zheng Zheng (2003) discussed the protection and update of peculiar streets[11]; and Yan Guotai (2000) tried to reveal the relationship between peculiar streets and the development of urban tourism[12]. The multi-dimensional and multi-level studies of domestic and foreign scholars on urban peculiar streets, provides necessary theoretical basis and practical experiences for our in-depth study on Shanghai leisure tourism peculiar streets.

Generally speaking, the meaning of urban leisure tourism peculiar streets includes three parts - "leisure tourism", "peculiar" and "streets". Leisure tourism is the basic function of peculiar streets, which means peculiar streets provide leisure tourism venue and service products for the public and other tourists through the attractiveness of its core products; peculiar is the theme, including both single and integrated forms, which is reflected in its theme performance elements; street is the form of its spatial existence, including two basic forms of narrow meaning of streets and broad meaning of blocks. The urban leisure tourism space units that have all the mentioned three features could be called peculiar streets.

Shanghai leisure tourism peculiar streets are the product of the development of Shanghai urban tourism. They rose around the earlier and middle 1990s, attracted more attention and concerns with the confirmation of the development targets of Shanghai urban tourism, and gradually entered the building and development stage. Of course, due to the differences in the initial stage and the maturity level of the market, most streets are still under construction or need to be improved, with the exception of only a few mature peculiar streets as Nanjing Road and Xintiandi etc. Nevertheless, it has become a fact that peculiar streets have been a tremendous impetus in pushing forward the development of Shanghai urban tourism industry [13]. At present, tens of fashion and consumer-oriented peculiar street blocks could be found in downtown and suburb of Shanghai, which are the concentrated expression of Shanghai local flavor and characteristics, have a certain attractiveness in market and product knowledge, with such leisure venues as restaurants, culture, entertainment, shopping malls, and comparatively comprehensive basic services and facilities.

From the perspective of the generation and development of peculiar streets, Shanghai peculiar streets could be generally summed as the following five types: First, peculiar streets formed through taking advantage of humanistic construction outlook in old urban blocks, such as the Yuyuan Garden and Chenghuang Temple Street. Second, peculiar streets updated and built on the basis of existing commercial districts, such as Nanjing Road Pedestrian Street, which evolved on the basis of original Chinese First Commercial Street. Third, peculiar streets which were formed with the combination of historical and cultural advantages and the rebuild of downtown areas, such as Xintiandi, a brand new peculiar street which was constructed in the rebuilding of Shanghai old Shi Kumen. Fourth, peculiar streets were built to make new functional position and regional development of residence area in suburb, in the adjustment of urban living population spatial structure, such as Meichuan Road in Putuo District. Fifth, peculiar streets formed by re-using old buildings such as old factories and warehouses in the new round of industrial restructuring of service industry in Shanghai, such as No. 8 Bridge, Tian Zi Fang and other cultural and creative industry Park.

Various forms of peculiar streets not only provide daily recreational service to the public, create a strong atmosphere of urban leisure, and effectively promote the construction of a harmonious living environment in the city, but also continuously enrich the connotation of urban tourism industry in Shanghai, expand the development space of urban tourism, and show development theory and development path for "Peculiar streets make a better life in urban leisure tourism" from another respect.

3. Spatial structure of peculiar street

The formation and development of Shanghai peculiar streets is a growing and spiral process. After more than 10 years of construction and development, there are 66 peculiar streets with various types, which have initially formed a network landscape structure and gradient and extended space which is complementary in function, dependant on each other in market, and with different characteristics in sight.

3.1 Spatial structure

3.1.1 The hierarchical structure

According to such market factors as market visibility, tourist receive flow and the attractiveness of services, peculiar streets could be divided into three levels[①]: first tier is peculiar street which has not only a local, but also a national and even international, such as Xintiandi, East Nanjing Road pedestrian street, and the Bund, 10 streets in total. The second tier is block with a certain market influence in Shanghai area, such as Yandang Road, Taikang Road blocks, also 10 streets in total. The third tier is peculiar street that has begun to take shape, but subject to various factors, their market impact is still relatively limited impact on the characteristics Street, a total of 46 (Table 1).

① Three hierarchies are based on the three rounds' result of the selection (recommendation, experts' election and the public to vote).

Table 1   The hierarchical structure of peculiar streets in Shanghai

Programme	Number	Percentage
The first hierarchical	10	15%
The second hierarchical	10	15%
The third hierarchial	46	70%
Total	66	100%

3.1.2   *The spatial structure*

In terms of the spatial structure of 66 leisure tourism peculiar streets in Shanghai, 47 of them are centralized in Luwan, Huangpu, Xuhui, Jing'an, Hongkou and Zhabei district, et al, about 71.2% of the total samples, and 19 are in other districts, about 28.8% of the total samples (table 2).

Table 2   The spatial structure of peculiar streets in Shanghai

District	Luwan	Huangpu	Xuhui	Hongkou	Jing'an	Zhabei
Number	12	11	6	5	5	3
Percentage	18.2%	16.7%	9.1%	7.6%	7.6%	4.5%

District	Yangpu	Putuo	Changning	Others	Total
Number	2	2	1	19	66
Percentage	3%	3%	1.5%	28.8%	100%

3.2   Character of spatial structure

3.2.1   *Concentration of the traditional core region*

The first character of Shanghai city peculiar streets' spatial structure is concentration of the traditional core region. The traditional core region for Shanghai means public and French colonies before liberation, including the whole Nanjing road, Huaihai road and Sichuan road[①]. There are 37 peculiar streets centralizing in this region, about 56.1% of the total samples (table 3).

Among 37 leisure tourism peculiar streets, 23 are centralized in an approximately rectangle area with Yan'an road[②] as axes and 2000 meters breadth from north to south (north to boundary of Nanjing Road, while south to boundary of Fuxing Road). From distribution denseness of city's central peculiar

---

①   Details reference: Zheng Zu'an. Memo of the name of place in old Shanghai [M]. Shanghai: Shanghai social sciencepress,1988.13-14(in chinese). 详参 : 郑祖安 :《上海地名小志》[M], 上海 : 上海社会科学出版社 ,1988 年 10 月 , 第一版 , 第 13~14 页。

②   Yan'an road includes east Yan'an road and west Yan'an road.

Table 3   The distribution of 66 peculiar streets in Shanghai

District	Number	Number (A)	Percentage (A)	Number (B)	Percentage (B)	Number (C)	Percentage (C)
Luwan	12	12	18.2%	10	15.2%	10	27%
Huangpu	11	8	12.1%	8	12.1%	8	21.6%
Jing'an	5	5	7.6%	4	6.1%	4	10.8%
Zhabei	3	1	1.5%	0	—	0	—
Xuhui	6	5	7.6%	1	1.5%	1	2.7%
Hongkou	5	5	7.6%	0	—	0	—
Yangpu	2	0	—	0	—	0	—
Changning	1	0	—	0	—	0	—
Putuo	2	1	1.5%	0	—	0	—
Pudong	4	0	—	3	4.5%	0	—
Minhang	2	0	—	0	—	0	—
Jiading	3	0	—	0	—	0	—
Fengxian	1	0	—	0	—	0	—
Jinshan	1	0	—	0	—	0	—
Songjiang	1	0	—	0	—	0	—
Nanhui	3	0	—	0	—	0	—
Qingpu	3	0	—	0	—	0	—
Chongming	1	0	—	0	—	0	—
Total	66	37	56.1%	26	39.4%	23	62.2%

Note: A means the peculiar streets in Shanghai public colony and French colony in 1930s; B means the peculiar streets in modern service industry district along Yan'an road; C means the peculiar streets that lie in not only colony but also modern service industry district along Yan'an road.

streets, it can be seen clearly that the denseness decreases in file from Yan'an Road to either north or south.

3.2.2 *Directivity of city CED*[①]

The second character of Shanghai leisure tourism peculiar streets lies in that it presents city CED guild state characterized by convergent regulation with city CED as its center. And the nearer CED, the more obvious it is, mainly attributing to magnet effect of city entertainment district. As early as 1930s, 10

---

① CED means Central Entertainment District with various functions of providing food in restaurants, fun in amusement sites, keeping fit in gyms and shopping in brick-and- mortar malls.

city entertainment districts with comparative amusement industry scale emerged in shanghai, including 1 CED and 9 sub-CED. Along with city modernization, these 10 city entertainment districts are witnessing different development levels with CED and nearer ones much more maturely as ever before.

Take sub-central amusement area—Huaihai road, south Shanxi road and south Maoming road for example. There are 7 leisure tourism peculiar streets, namely Huaihai road, south Maoming road, Gaolan road, Xiangshan road, Sinan road, Changle road and middle Fuxing road within radius of 500 meters, which can be concluded as high convergence rate, exceeding average level of 6.6. While the convergence rate is much higher in central amusement area, by contrast, that in sub-central amusement area such as old west temple one far away from it is relatively weak.

3.2.3 Inosculation with modern service industry development axes

East and West Zonal convergent rule with Yan'an Road—Century Avenue Service industry as axes constitutes its third character. With reference to rectangle range of 2000 meters breadth (north to boundary of Nanjing road, south to boundary of Fuxing road, east to boundary of Century Avenue), 39.4% of total samples as table 3 shows, are centralized in this long and narrow area. If seen from transverse angle with Huangpu river as its boundary, namely from Yan'an road peculiar streets in Puxi is more dense than that in Pudong district. However, if seen from vertical angle, it is clear that the closer to axes, the more dense its convergence. Such rule highly matches the principle of developing Shanghai modern service industry that it will exert its utmost to construct three service industry convergent band by making good use of area superiority as well as city function conversion and meanwhile, give high priority to build up Huangpu river, Suzhou river—the modern service convergent area where it reveals charm of metropolis' function and image and Shanghai Business Aisle--Yan'an Road with Century Avenue as its axes [14].

3.2.4 The coherence of city space expansion

The forth character in terms of Shanghai peculiar streets' spatial framework lies in the coherence of city space expansion, namely convergent distribution regulation paralleled with east north and west south of Huangpu river, mainly due to its city evolvement trace characterized by extension from areas like Nanjing road and people's square to northeast and southwest direction. There exist 36 leisure tourism peculiar streets, including 2 in Yangpu district, 5 in Hongkou district, 11 in Huangpu district, 12 in Luwan district, and 6 in Xuhui district, which as a whole account for 54.5% of total numbers of such peculiar street.

4. The illustration of multi-fold cause

The emergence, development and prosperity of Shanghai peculiar streets are no accident, but are closely related to the fate of the city, including its profound cultural background, modern city functions change and future development trends.

4.1 Perfect combination of cultural legacy and modern elements

Firstly, Shanghai has developed as a metropolis characterized by multi-culture, owing to its former colonized culture accumulated for hundred years and latter assimilation of Chinese civilization. Built up as the International public colony and French colony site, Shanghai has been open to outside world with merchandize pouring in followed by Occidental physical civilization, politics regime and mental thoughts at an unparalleled rate (Zheng Zu'an, 1999) [15]. In terms of physical aspect, its main task is to improve city's landscape by constructing elegant arc architecture group in Bund, prosperous view of colorful Nanjing road, unique residential skyscrapers, sequestered novel lane as well as garden villa, and extraordinary beautiful Xiafei road (current Huaihai road) which were once ranked as highly booming symbol of public and French colony site. Besides there yet existed rows of Shi Kumen in Shanghai. On the other hand, considering its mental aspect, Shanghai dwellers have already led a novel life, owing to penetrated life idea, city construction experience and management mode that are superior to rest cities at home.

Secondly, the construction mode of leisure tourism peculiar street, as a combination of modern life elements and cultural relics, not only fully tap the soul of a city as well as enhance its value, but also bring about a practical and referential significance regarding how to protect and exploit it in a novel way. In the pursuit of personality as a fashion, owing to precious cultural legacy, Shanghai is endowed with rich material including large quantities of excellent but classic architects that are conducive to characteristic build of this city, what's more, its perfect combination with modern fashionable elements also give rise to a classic. Take successful development of Xin Tiandi and Bund as a potent proof, which has already been granted as city cards, thus gradually building up a brand.

4.2 Important carrier of modern tourism development

The position as a metropolis of Shanghai requires that it should exert its utmost to develop modern tourism. Generally speaking, metropolis must be a developed tourism city, whereas it is not necessary (Peking International City Development Institute, 2002) [16]. As it is well known, modern tourism plays a more and more important role in city's function, while the lack of related resources has been a bottleneck for its modern tourism development. Therefore, former streets as a miniature of city development and historical elite can be regarded as a window for show, especially combined with more tourism elements and novel connotation, thus evolving to an increasingly indispensable carrier for modern tourism development. So that, both of them can advance mutually.

Firstly, to spark value of current old street resources by using modern tourism products and market thought. Take technical package and rebuild of bund's architectures full of hundred-year history and Nanjing Road for example, not only enable their opening time and space queen more desirable, their resources distribution more optimum, but also grant their view function from plane to solid, from one-fold to multi-fold, thus making the leisure tourism peculiar streets more specialized and mature, meanwhile, enhancing its role

Secondly, advanced development of modern tourism increasingly gives birth to new leisure tourism peculiar streets. For instance, owing to its unique style, Meichuan road built in 2004 full of European taste located in Putuo district not only enrich regional tourism resources, but also improve overall tourism image of this district. For another, unprecedented prosperity of Shanghai's modern tourism provides a god-given opportunity for its further development, and will be conductive to its quantity accumulation even quality leap as a result.

4.3   The necessity of leisure city life

114 days for holiday by law added by annual holidays and summer or winter holidays, do enable people living in Shanghai to enjoy leisure life for almost 1/3 of the whole year. According to international development experience, the threshold of overwhelming city life towards leisure is at such critical point when per GDP amounts up to between 3000 and 5000 dollars. The statistic shows that breaking through 5000 dollars per GDP for Shanghai city dwellers can be seen as a symbol of sufficient time and money spent on leisure, which means that Shanghai city has stepped into leisure lifestyle and that could be the tendency for its development in future.

Firstly, satisfying leisure need of white collar in business district should be taken high priority. Leisure is not only as a way for showing personality, perfecting human dignity and improving quality level, but also as an approach for doing business and social activities. Therefore, any leisure facilities and service for this group will stand for a high point concerning level of a city's leisure life. In this sense, top-grade peculiar streets with various functions of providing food in restaurants, fun in amusement sites, keeping fit in gyms and shopping in brick-and- mortar malls. Thus its coming forth is destined although by chance.

Secondly, popular leisure status requires that leisure tourism peculiar streets should be constructed multi-fold. Though in terms of such important index as leisure time and per GDP, Shanghai has already met the lowest standard well acknowledged by leisure theory, it yet seems a long way to go before reaching real leisure times concerning related factors as the way to enjoy leisure, overall quality of city dwellers and disposable incomes. In fact, its popular leisure is still at lower level. Therefore, leisure tourism peculiar street is coming forth. Meanwhile, such peculiar street characterized by unique design, elegant environment and considerate service, not only meet the current needs but also create new ones, thus leading cockle upward of leisure life standard, which is ranked as guidance for city leisure life.

5. Advice of improving the healthy development of peculiar streets

From the point of its current spatial layout and the cause of the multiplicity, the peculiar streets' successful exploitation depends on the combination and allocation of district, resources and market it has, as Hengshan Road Leisure Entertainment Street. There are not only many buildings of European garden style, but also some perfect service establishment like bars, bowling, and tennis centers along the way. Besides, its north is near the fashion shopping district-Huaihai Road; its south is connected with the business center-Xujiahui, where the demand is brisk; and the traffic is convenient. In the view of this, toward the full swing exploitation of districts and counties' construction of peculiar streets, the author made the following recommendations:

5.1 Seeking rational spatial layout

Spatial layout contains the spatial arrangement between streets and the streets' interior spatial distribution. Seeking rational spatial layout, one side we should pay attention to the balance of site selection between distinct and counties and its interior parts, trying to make the market complementary; the other side we should pay attention to the layout balance between the business units, avoiding unnecessary loss caused by competition, to realize the whole street's common prosperity sequentially. At the same time, we should deal with the relationship between the common area and private room to make the citizens and tourists enjoin the largest freedom and the best experience, in the case of not interfering the owners' normal business.

5.2 Give prominence to subjectify the function orientation

Subject is one manifestation of art and one creation of block culture in essence, and it can be abstracted and transformed also. Giving prominence to subjectify the function orientation is showing its characteristic. Therefore, when we make the peculiar streets' function orientation, we should balance its connotation, advantages and abstract its subject with the most symbolic features, embodying many details as block signs, products and so on to achieve our subjective of creating the blocks by the theme. The using of art can make the peculiar streets have a larger cohesive force to realize the maximal social economic efficiency.

5.3 Trying to look for the optimization of the resources distribution

The allocation of resources in different time, different district and its usage is called resources distribution (Lou Jiajun, 2001)[17]. Trying to look for the optimization of the resources distribution is to deal with relationship of the markets, transportation and the products supply. Especially we should deal with the outside entrance and inside circulation, ensuring the close intercourse of supplies and demands, to realize the utmost resource efficiency.

5.4 To realize the market-oriented whole operation

Nowadays, as one part of community building, peculiar streets are mainly invested by government. However, we should be cognizant that the government's behavior will be instead by the enterprise's investment in the future. First, the exploitation and construction is a kind of economic phenomenon, which is a part of city tourism, and it will realize the economic returns finally. Second, enterprise's investment can avoid some disadvantage caused by government's behavior, and it can mobilize the investors' positivity to make the peculiar streets healthy and sustainable development. Therefore, government should have a correct location and make good use of all social resources. Make each peculiar street withstand the test of market and offer better services to citizens and tourists to realize the real "Shanghai-exciting every day".

Thanks: Give much thanks to Mr. Li Ping who works in Shanghai Tourism Management Committee for his help with the samples, and Ms. Li Mei, Ms. Duan Jisheng, Ms. Zhu Ningning for their help with the translations.

References

［1］Einar Lillebye. Architectural and functional relationships in street planning:an historical view［J］. *Landscape and Urban Plan*,1996, 35:85–105.

［2］Assenna Todorova, Shoichiro Asakawa, Tetsuya Aikoh. Preferences for and attitudes towards street flowers and trees in Sapporo, Japan［J］. *Landscape and Urban Plan*, 2004,69:403–416.

［3］Stephen W Litvin. Streetscape improvements in an historic tourist city a second visit to King Street, Charleston, South Carolina［J］. *Tourism Management*, 2005, 26:421–429.

［4］Tang Fanming. General Introduction of the Study of the Position and Role Which Pedestrain Shopping Streets Have and Play in Tourism Industry［J］. *Journal of Guilin Institute of Tourism*, 2005, 16(6):83–85(in Chinese). 唐凡茗：《步行街在旅游业中的地位和作用研究概述》［J］，《桂林旅游高等专科学校学报》，2005 年第 6 期，第 83~85 页。

［5］Zheng Xiaoshan. An Analysis of the Design of Chunxi Road Commercial Pedestrian Mall Based on Place Theory［J］. *Journal of Southwest Jiaotong University(Social Sciences)*, 2004, 5(6):139–142(in Chinese). 郑晓山：《基于场所理论的 jiaosheng 春熙路商业步行街设计分析》［J］，《西南交通大学学报》(社会科学版)，2004 年第 6 期，第 139~142 页。

［6］Jiao Sheng,Ceng Guangming,Li Xianghui,et al.Study on the Compound Mode of Development of Commercial Pedestrian Street in Old City［J］. *Building Science Research of Sichuan*, 2005, 31(5): 118–121(in Chinese). 焦胜、曾光明、李向辉等：《旧城商业步行街的复合开发模式探讨》［J］，《四川建筑科学研究》，2005 年第 5 期，第 118~121 页。

［7］Tang Yue. On the Relation of Specialty Shopping Streets with Tourism［J］. *Journal of Zhejiang Shuren University*, 2004, 4(3): 28–30(in Chinese). 唐跃：《略论商业特色街与旅游的关系》［J］，《浙江树人大学学报》，2004 年第 3 期，第 28~30 页。

［8］Bao Jigangetal. *City Tourism:Theory and Case*［M］.Tianjin: Nankai University Press, 2005.160(in Chinese). 保继刚等：《城市旅游：原理 · 案例》［M］，天津：南开出版社，2005，第 160 页。

［9］Yao Kunyi. On the Location of Henshan Tourism Street in Shanghai［J］. *Tourism Science*, 2000(1): 9–11(in Chinese). 姚昆遗：《关于上海衡山路旅游街定位的思考》［J］，《旅游科学》2000 年第 1 期，第 9~11 页。

［10］Chen Jikun.On the Function Improvent of Fuzhou Culture Street［J］. *Shanghai Statistics*, 2001(04):33–34(in Chinese). 陈继昆：《浅谈福州路文化街功能的完善与提升》［J］，《上海统计》2001 年第 4 期，第 33~34 页。

［11］Zheng Zheng. Preservation and Renewal Planning of Duolun Road Pedestrian Precinct ［J］. *Urban Planning Forum*, 2003, 147(5): 34–40(in Chinese). 郑正：《上海市多伦路商业文化休闲步行街区保护更新规划》［J］，《城市规划汇刊》，2003 年第 5 期，

第 34~40 页。

[12] Yan Guotai. Quality of Leisure Street and Urban Tourism［J］. *Journal of Tongji University Social Science Section*, 2000, 11(1):29–33(in Chinese). 严国泰:《休闲街的品位与都市旅游》[J]，2000 年 11 月第 1 期《同济大学学报》(社会科学版)，第 29~33 页。

[13] Song Changhai.Innovative Development of Urban Tourism Resources in Shanghai［N］. *China Tourism News*, 2006–03–17(5)(in Chinese). 宋长海:《上海创新开发都市旅游资源》[N]，2006 年 3 月 17 日第 5 期《中国旅游报》。

[14] Shanghai Modern Service Industry built three Bands and nineteen Districts［N］. *Wenhui Daily*, 2005–02–23(5)(in Chinese).《申城现代服务业构筑 "3 带 19 区"》[N]，2005 年 2 月 23 日第 5 期《文汇报》。

[15] Zheng Zu'an. *One Hundred Years of Shang City*［M］.Shanghai: Xuelin Press, 1999.3(in Chinese). 郑祖安:《百年上海城》[M]，上海:学林出版社，1999 年 4 月第 1 版:前言第 3 页。

[16] Peking international city development institute. Figures of China–Non–confidential figures reading of China［M］. Guangming daily press, 2002.313(in Chinese). 北京国际城市发展研究院:《数字中国 – 中国非保密性数字读本》[M]，光明日报社，2002 年出版，第 313 页。

[17] Lou Jiajun. On the Impact of Tourist Resource Allocation on the Development of Shanghai's Urban Tourism［J］. *Tourism Tribune*, 2001, 16(2): 56(in Chinese). 楼嘉军:《试论资源配置对上海都市旅游发展的影响》[J]，《旅游学刊》，2001 年第 2 期，第 56 页。

A Study on Industry Cycle Index of China's Travel Services Since 1993

Dai Bin, Yan Xia, Huang Xuan

(School of Tourism Management, Beijing International Studies University, Beijing 100024, China)

Abstract: The paper constructs an industry cycle index model after a systematic arrangement of statistical data since 1993. Based on quantitative study, it makes a comment on the industrial contribution rate in the process of its development.

Key words: travel service industry; business cycle; cycle index

1. Introduction

In 1940's, National Bureau of Economic Research (America) advanced a method calculating the macroeconomic cycle using mutiple index instead of a single index [1]. Since then, the method was used to study the developing situation of all industries [2]. With the accumulative knowledge of prosperity cycle and the development of econometric techniques, many new methods such as growth cycle method, principal component analysis, state space model, were starting to be used in many countries for the prosperity analysis. Despite some prosperity index research such as Report on China Industry Prosperity Analysis, China

[Fund] This paper is one of the main reports for the project of National Natural Science Foundation entitled "A Study on Industry Cycle Index of China's Tourism" (No. 70640012). The first draft is written by Yan Xia.

[About authors] Mr. Dai Bin, professor of Tourism at Beijing International Studies University, with research interests centred on issues related to tourism enterprises management and economy of tourism industry; Ms. Yan Xia, a graduate student at School of Tourism Management in BISU; Mr. Huang Xuan, a graduate student at School of Tourism Management in BISU.

Economic Index, Enterprise Prosperity Survey, they are, however, done in the macroeconomic perspective and limited to only a few industries. Recent years has seen the emergence of cycle research of the tourism industry in the form of books and special reports, but there is no systematic study of the prosperity index of travel services on the basis of model building. At present, the theoretical absence of macro-monitoring of the travel services and long-term prediction in our country greatly harms the capability of industrial regulation on the part of governments and the capability of strategic judgement on the part of enterprises.

2. Research methodology and model building

Industry Prosperity Indexes of Travel services in this paper refers to cycle model on the basis of statistics which is used to yield quantative description and qualitative judgements of the change in the travel services industry, reflecting the status quo and the future trend of the economic performance of the travel services industry and the production and operation of enterprises. This paper tries to establish the Industry Prosperity Index of China's Travel Services with a view to developing the research paradigm in this field through various methods including mathmatic model building, data collection and preprocessing, index selection, statistical analysis and qualitative methods[3].

2.1 Data source and index selection

2.1.1 Data source
Data used in this paper are mainly from the statistics publicly available from China National Tourism Administration (CNTA), such as *the Yearbook of China Tourism Statistics, the Yearbook of China Tourism Statistics (Supplement), the Yearbook of China Tourism*. The statistics of total output of the third industry and GDP come from the analysing system of TianXiang Investment Consulting Company.

2.1.2 Index selection and base period determination
In index selection, principles of respect for the experts' opinions and the integrity of index data are observed. After analysing the correlation of the 60-odd statiscal indexes of the travel services from CNTA and considering the consistency and comparability of the statistics, we eliminate the indexes which are either incomplete, inconsistent in statistic aperture,or having a high correlation. The inspection of travel service of category 1 and category 2 began in 1991, and the inspection of travel service of category 3 began in 1993, considering the data before 1990 were both incomplete and incompatible, so 1993 is decided as the

base period. Finally, the Enterprise Scale Prosperity Index, Enterprise Operation Prosperity Index, Market Prosperity Index and Annual Inspection Prosperity Index are decided as the four sub-indexes. Among these, Enterprise Scale Prosperity Index includes Enterprise Number Prosperity Index and Employee Number Prosperity Index; Enterprise Operation Prosperity Index includes revenue, profits, profitability, tax, assets, and productivity; Market Prosperity Index includes inbound tourism, domestic tourism, and outbound tourism, and each Market Prosperity Index is comprised of Liaised Person-time, Received Person-time and Marketization Rate. Annual Inspection Prosperity Index refers to the passage rate of travel servives in their annual inspection.

2.1.3 Data preprocessing

Regulations on Administration of Travel Services promgulated in 1996 divided travel services into international and domestic ones, instead of the previous 3 catogories. Therefore, the statistics of international travel services before 1996 are the sum of travel services of category 1 and category 2, and statistics of demostic travel services before 1996 are those of travel services of category 3. However, some statistics are missing despite our intense efforts, and are dealt with as follows: the missing passage rate of both demestic and international travel services for the year 1996, 2003 and 2004 will take the average passage rate of the rest years as referecne; and for the missing 3-year demestic organized person-time, we first analyze the average, the growth rate and the ratio of demestic organized person-time to domestic Received Person-Time,and the statistics of the rest years will be taken as reference, finally, the demestic organized person-time will be determined on the basis of same-year domestic Received Person-Time.

2.2 Weight determination

In the process of research, we consulted many experts, scholars and people from the business circle for their opinions on weight determination, and after careful study and consideration, we finally decided upon the most objective method in statistics—variation coefficient method, by which the weight of indexes are determined utilising the information contained within the indexes. First, an index evaluation system is established. It's supposed to include i (i=1, 2, ..., n) indexes, and j (j=1, 2, ..., n) object will be evaluated. The actual values of the indexes of objects are supposed to be used to calculate the averages and

standard deviation of indexes. The average of index i, $\bar{x}_i = \dfrac{\sum\limits_{j=1}^{m} x_{ij}}{m}$;the standard deviation of index i, $\sigma_1 = \sqrt{\dfrac{\sum\limits_{j=1}^{m}(x_{ij}-\bar{x}_i)^2}{m}}$; then variation cofficient is calculated,

$V_i = \dfrac{\sigma_i}{\bar{x}_i}$ (i=1, 2, ..., n); Finally, the weight of all indexes are calculated, $W_i = \dfrac{V_i}{\sum\limits_{i=1}^{n}V_i}$.

2.3   Index calculation

With reference to the calculating method of prosperity indexes in other industries, especially in national economy, and considering the features of the travel services, indexes are calculated as follows: for example, under the 3-level prosperity index system, with 1993 as the base year, the 4th-level index n under 3th-level index i in year m would be calculated: $P_{\min} = \dfrac{X_{\min}}{X_{1993in}} \times 100\%$ (m stands for the year; i stands for the 3rd-level index i ;n stands for a 4th-level index, i.e., the in 4th-level index). After that, on the basis of the decided weight, all the 3rd-level prosperity indexes in every year will be calculated,e.g.,the 3rd-level index i in year m would be: $P_{mi} = P_{mi1}W_{i1} + P_{mi2}W_{i2} + K + P_{min}$(n stands for the number of sub-indexes of index i). Likewise, the 2nd-level prosperity index would be achieved on the basis of the 3rd-level indexes and their weight. And eventually Prosperity Index of International Travel Services, Prosperity Index of Domestic Travel Services and Industry Prosperity Index of China Travel Service will be calculated.

2.4   Research methodology and its verification

Through mathematic calculation, Prosperity Index of International Travel Services, Prosperity Index of Domestic Travel Services and Industry Prosperity Index of China Travel Service can be calculated respectively, and then the ratio of international travel services to domestic travel services is to be achieved using the coefficient variation method. Therefore, Industry Prosperity Index of China Travel Service is available and it can be compared with the one that is directly calculated with the statistics so as to verify the effectiveness of the research methodology.

3. Result analysis of prosperity indexes

3.1   Prosperity index of international travel services

3.1.1   The enterprise scale prosperity index of international travel services

The Enterprise Number Prosperity Index of International Travel Services can be divided into two stages: The initial stage is from 1993 to 1999, in which the number fluctuated but grew on the whole; and it has entered into the stage of steady growth since the 21st century. Taking the impact of SARS in 2003 out of consideration, Enterprise Number Prosperity Index of International Travel Services experienced similar growth. The difference is that international travel services didn't experience obvious growth in the initial stage but the growth in the second stage was magnificent; while Employee Number Prosperity Index experienced a fast growth in the intial stage but it slowed down in the second stage. The Enterprise Scale Prosperity Index of international travel services experienced similarly and grew on the whole, with its summits in 1995, 1998 and 2003. It can be predicted that in the following years, The Enterprise Scale Prosperity Index of International Travel Services will continue to rise but slowly a little bit.

3.1.2 Enterprise operation prosperity index of international travel services

Enterprise Operation Prosperity Index is determined by six indexes including revenue, profits, profitability, tax, assets, and productivity. All these indexes went upwards except profitability. Similar to The Enterprise Scale Prosperity Index of International Travel Services, Enterprise Operation Prosperity Index made a clear-cut division before and after the year 1999: the last decade of past century saw steady growth of all other indexes, among which productivity and revenue kept growing fast after 1995 while assets, tax and profits grew at a slow pace. Except productivity, most indexes entered into the new century with high growth rate and remained prosperous in the beginning of the century, and they were affected by SARS in 2003 to different extents and recovered in 2004. On the whole, Enterprise Operation Prosperity Index kept a steady growth by the end of the past century and has experienced faster and good growth since 2000 (except 2003). According to the current situation, it will remain prosperous in the following 2-3 years.

3.1.3 Market prosperity index of international travel services

Market Prosperity Index of International Travel Services is mainly determined by five indexes including Inbound Liaised, Inbound Received, Inbound Mercerization Rate, Outbound Organized, and Outbound Marketization Rate (the statistics of domestic market of international travel services is missing). International travel services are prosperous in outbound organized market, despite temporary downturn after its summit in 2000, and it kept growing after that. But relatively speaking, there was slow growth in Outbound Marketization Rate by the end of the last century and the rate was going downward pretty fåst in this new century. The inbound tourism market reached its lowest point in 1998,and then it began to grow fast and reached its summit in 2002. It was affected by SARS in 2003 and recovered afterwards. The prosperity level of Inbound Received was a bit higher that of Inbound Liaised and the gap widened after 1998. Inbound Marketization Rate continued at the same trend as the Inbound Liaised and Inbound Received, but the property level was lagging far behind the other two and it has seen no improvement since 2003. Overall, Market Prosperity Index of international travel services grew at a slow rate before 1998, and began to grow fast thereafter (except in 2003), and this fine growth trend will not change much in the following 3 or 5 years.

3.1.4   Annual Inspection Prosperity Index of International Travel Services

Annual Inspection Prosperity Index of International Travel Services fluctuated, with its low period between 1998 and 2001, and it has been stable in recent years. The index is not expected to experience much growth in the coming years considering the intermittent promgulation of regulations on the operations and management of travel services.

Figure 1 illustrates the changes of Prosperity Index of International Travel Services after comprehensive compilation.

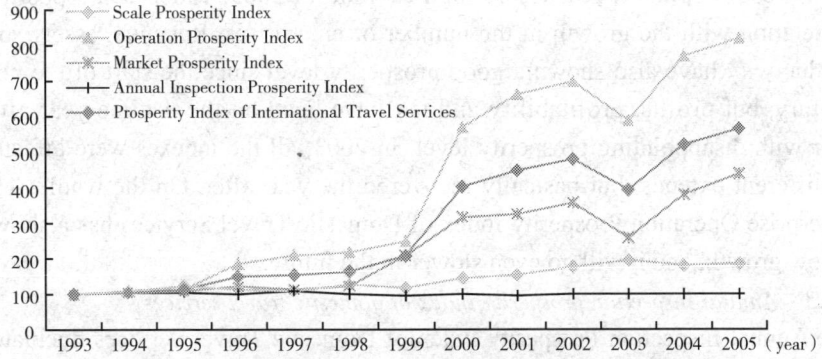

Figure 1   Prosperity Index of International Travel Services

Source: *the Yearbook of China Tourism Statistics, the Yearbook of China Tourism Statistics (Supplement), the Yearbook of China Tourism (1994-2006)*

The above-mentioned four sub-prosperity indexes kept the same trend, with Enterprise Operation Prosperity Index the highest, followed by Market Prosperity Index, and Annual Inspection Prosperity Index grew at the lowest rate. The Enterprise Scale Prosperity Index and the Annual Inspection Prosperity Index were not affected by SARS, but the other two indexes met great frustrations. From the stages of the historical process, all indexes underwent relatively slow growth before 1999, but grew faster in the new century. In general, indexes of the international travel services were good, especially after 1999(2003 excluded). In the coming years, the international travel services are going to continue prosperous trend.

3.2 The prosperity index of domestic travel services

3.2.1 The enterprise scale prosperity index of domestic travel services
The Prosperity Index of Domestic Travel Services has been enjoying robust growth both in the number of enterprises and that of employees. Relatively speaking, the past couple of years have seen gentle growth of employees due to factors like the quality of employees and scientific techniques. While the growth in the number of enterprises will not necessarily slow down, The Enterprise Scale Prosperity Index of Domestic Travel Services will to a certain extent.

3.2.2 The enterprise operation prosperity index of domestic travel services
In general, The Enterprise Operation Prosperity Index of Domestic Travel Services has been on the upward trend at an increasingly slower pace year after year. It can be divided in two stages: It didn't grow fast on the whole before 1999,but after that it quickened its pace. The revenue property index grew at the fastest rate, and it reached as high as 4000 in 2005, which had a positive correlation with the growth in the number of enterprises. Revenue, assets, and productivity have also shown a good prosperity level since the start of the new century, but profits, profitability and tax have been on the decline year after year with disappointing prosperity level. In 2003, all the indexes were affected to different extends, but basically recovered the year after. On the whole, The Enterprise Operation Prosperity Index of Domestic Travel Services has a shown a slow growth, and it will go even slower in the future.

3.2.3 Annual inspection prosperity index of domestic travel services
The Annual Inspection Prosperity Index of Domestic Travel Services fluctuated but on the whole it showed a good prosperity level, with a steady or even declining trend.

Figure 2 illustrates the changes of Annual Inspection Prosperity Index of Domestic Travel Services after comprehensive compilation.

The Enterprises Scale Prosperity Index of Domestic Travel Services went up at a constant pace and maintained a fast growth; Influnced by many factors, Enterprise Operation Prosperity Index was going up but fluctuating a lot; Annual Inspection Prosperity Index remained a very slow growth rate. All in all, Prosperity Index of Domestic Travel Services grew very fast in last century, but it slowed down in the new century, and it will grow even more slowly in the coming years.

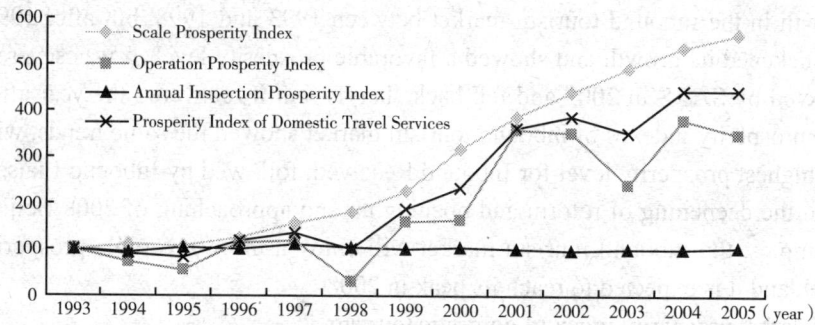

Figure 2   Prosperity Index of Domestic Travel Services
Source: *the Yearbook of China Tourism Statistics, the Yearbook of China Tourism Statistics (Supplement), the Yearbook of China Tourism (1994-2006)*

3.3 Prosperity index of travel services industry

3.3.1 *Scale prosperity index of travel services*
Looking from the enterprises scales, the whole travel services industry maintained a good growth rate whether measured by the number of enterprises or by that of employees. It suffered its first magnificent fallback in 2004, but it recovered within less than a year. Enterprise Number Prosperity Indexes showed a better prosperity level than of Employee Number Prosperity Index, and the gap will continue to widen. All in all, the good prosperity level of the whole industry will continue, but with an increasingly slower pace.

3.3.2 *Operation prosperity index of travel services industry*
Revenue Prosperity Index of the whole industry maintained a good growth rate; Assets Prosperity Index grew at a slow pace except for the years 1999 and 2000;Tax Prosperity Index grew slowly and reached its summit in 2001,followed by a fallback,with its lowest ebb in 2003 due to SARS,and recovered afterwards; Productivity Prosperity Index grew fast before 1997 and then it grew slowly; Profits Prosperity Index and Profitability Prosperity Index suffered negative growth basically, and especially, Profitability Prosperity Index was negative in rencent years. In 2003, all indexes were affected by SARS to different extents. On the whole, the Operation Prosperity Index of the whole industry maintained a steady growth rate except some leaps in 1999 and 2000. The trend is going to continue in the years to come.

3.3.3 *Market prosperity index of travel services industry*
(1) Prosperity index of inbound tourism
The whole industry showed unfavorable prosperity level and suffered negative

growth in the inbound tourism market between 1993 and 1998, but after 1998, it quickened its growth and showed a favorable prosperity level. It was severely affected by SARS in 2003 and fell back sharply, but it recovered the year after. The prosperity indexes of inbound tourism market showed the same trends, with the highest prosperity level for Inbound Received, followed by Inbound Liaised. With the deepening of reform and opening up and approaching of 2008 Beijing Olympics, the inbound tourism market will enjoy a more favorable prosperity level, and it is expected to reach its peak in 2008.

(2) Market prosperity index of domestic tourism

Market Prosperity Index of Domestic Tourism maintained a good growth rate in general and it reached its peak in 1998.It fell back in 1999 and maintained a fast growth, and SARS in 2003 didn't affect its performance in 2004, but the growth slowed down. All prosperity indexes of domestic market showed the same growth, and among these, the Liaised of Domestic Tourism Market enjoys more favorable prosperity level than Received of Domestic Tourism Market. And the Marketization Rate remained steady in recent years and it was on the downward trend.

(3) Prosperity index of outbound tourism market

Prosperity Index of Outbound Tourism Market enjoyed sustained growth,espcecially during the period between 1997 and 2000. And it started to fall back in the first couple of years of the new century. After SARS in 2003, it quickened its growth and the trend will continue. However, Marketization Rate remained steady in the last century and it went downwards after 2000. Generally speaking, the outbound tourism market had becoming increasingly prosperous, and it will demostrate greater prospeirty level with the development of China's economy and the opening of outbound tourism destinations for Chinese people.

The Prosperity Index of Outbound Tourism Market kept growing pretty fast, and it fell back after it reached 488 in 2000. But the growth was ever more robust after SARS, with its peak 709 in 2005. The domestic market was the most prosperous despite its strong fluctuations, its prosperity index was 530 in 1998, and it reached 1002 in 2004.Relatively, the Prosperity Index of Inbound Toursim Market grew at the slowest rate, and wandered between 70 and 80. It rose pretty fast from 163 in 1999 to 331 in 2002, and it went down to 174 in 2003, followed by fast growth again. All in all, the Market Prosperity Index looks promising and it will go gentle.

3.3.4 Annual inspection prosperity index of travel services industry

Annual Inspection Prosperity Index demostrated strong fluctuations between 1993 and 1998 due to the instability of institutional environment and market environment. It will become more stable with the improvement of the institutions and standardization of the industry.

Figure 3 illustrates the changes of Prosperity Index of Travel Services Industry after comprehensive compilation.

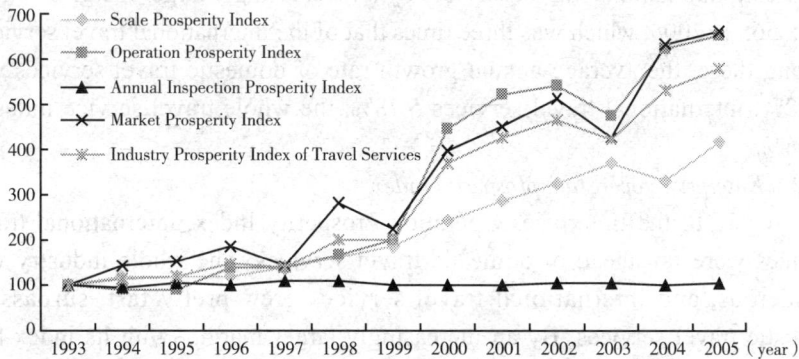

Figure 3   Prosperity Index of Travel Services Industry in China (1993-2005)

Source: *the Yearbook of China Tourism Statistics, the Yearbook of China Tourism Statistics (Supplement), the Yearbook of China Tourism (1994-2006)*

The Prosperity Index of Travel Services Industry and its sub-indexes have been good. In 1999, all indexes except Annual Inspection Index were around 200 and 1999 became the watershed for the sub-indexes. During the period between 1993 and 1999, Market Prosperity Index was the highest, and Enterprises Operation Prosperity Index took its place after 1999. In the past 13 years, Enterprises Operation Prosperity Index and Market Prosperity Index have surpassed the Enterprises Scale Prosperity Index by increasingly larger margin, which implies that the extention of the overall supply by travel services belong to the external economy. In 2003, all indexes but Annual Inspection Index were affected by SARS, which verifies the proposition that the travel services industry is a sensitive and vulnerable one. Prosperity Index of China Travel Services has been keeping steady growth in the past 13 years. The trend is promising and will continue to go up with the development and maturation of the industry.

4. Comparative analysis and verification of research methodology

4.1 Comparison among international travel services, domestic travel services and the travel services industry

4.1.1 Enterprises scale prosperity index
With regard to Enterprises Scale Prosperity Index, domestic travel services

were far ahead of international travel services. According to the indexes, the whole industry was prosperous, and domestic travel services grew pretty fast, surpassing international travel services by an increasingly larger margin, with its index 565 in 2005, which was three times that of the international travel services. Among these, the average annual growth rate of domestic travel services was 16.02%, international travel services 5.78%, the whole travel service industry 13.06%.

4.1.2 Enterprises operation prosperity index

With regard to the Enterpises Operation Prosperity Index, international travel services were far ahead of domestic travel services. The whole industry was prosperous, and international travel services grew pretty fast, surpassing domestic travel services. By an increasingly larger margin, with its index 819 in 2005,which was twice that of the domestic travel services. Among these, the average annual growth rate of domestic travel services is -15.16%, international travel services 23.41%, the whole travel service industry 20.83%.

4.1.3 Market prosperity index

With regard to Market Prosperity Index, the international travel services are below the whole industry. The growth rate of the Market Prosperity Index of international travel services was 16%, which was below that of the whole industry, 22.14%, and their indexes were 440 and 660 respcetively in 2005.

4.1.4 Annual inspection prosperity index

With regard to Annual Inspection Prosperity Index, the domestic travel services were relatively higher than the international travel services and the whole travel service industry, and have gone gentle in recent years. The average annual growth rate of prosperity index for the domestic travel services was 0.66%, the international travel services 0.08%, the whole industry 0.47%.

4.1.5 Industry prosperity index

With regard to Industry Prosperity Index, international travel services are much higher than domestic travel services, and slightly higher than the whole travel services industry. The average growth rate of Property Index for international travel services is 17.87%, domestic travel services 16.88%, and the whole travel services industry 17.92%.

There was law in the fluctuation of the Prosperity Index of Travel Services Industry: The average fluctuation rate was relatively low at 13.79% between 1993 and 1999, and it was much higher at 24.3% between 1999 and 2003.It went down a little bit to 17.56% after 2003. Therefore, the development of travel services industry is divided into three cycles, and the third cycle will last at least till 2010 if not strongly affected by external factors like unexpected events or

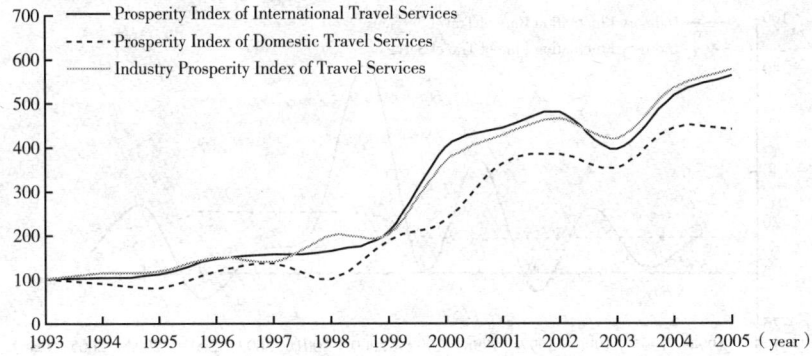

**Figure 4   Comparison of Prosperity Index of Travel Services
Industry (1993-2005)**
Source: *the Yearbook of China Tourism Statistics, the Yearbook of China
Tourism Statistics (Supplement), the Yearbook of China Tourism (1994-2006)*

financial crisis.

Taking the average fluctuation rate of Industry Prosperity Line of each cycle as the annual fluctuation rate of each cycle, a prosperity line of travel services industry can be drawn. If the prosperity index falls into the left upper side of the line, the industry is deemed to prosperous. And it's not if otherwise.

In the first cycle (1993-1999), Industry Prosperity Index fluctuated on an upward trend at a slow pace, having a low dispersion rate with the line. The second half of the cycle (1996-1999) had a higher fluctuation rate: 1997 experienced negative growth but 1998 saw the highest growth rate of 45%. It was moderately prosperous in 1994 and 1996, and the deviation with the prosperity line was around 2; it was not prosperous in 1995 and 1999, and the deviation rates were -11 and -14 respectively; it was not prosperous and the deviation was -29; it was prosperous in 1998 and the deviation was 10. In the second cycle (2000-2003), Industry Prosperity Index grew fast on the whole, but the fluctuation rate went downward year after year, and it was on the upper side of the line, demonstrating a high dispersion with the prosperity line. It was very prosperous between 2000 and 2002, and the deviaions were 99, 92, and 48 respectively. It was not prosperous in 2003, with the deviation at -97. Industry Prosperity Index grew at a slower pace on the whole in the third cycle (2004-) and it was below the prosperity line, demonstrating a high dispersion with the prosperity line. It was not prosperous in the 2004 and 2005, with deviations -76 and -139 respectively.

Looking across the 13 years, the Industry Prosperity Index of Travel Services

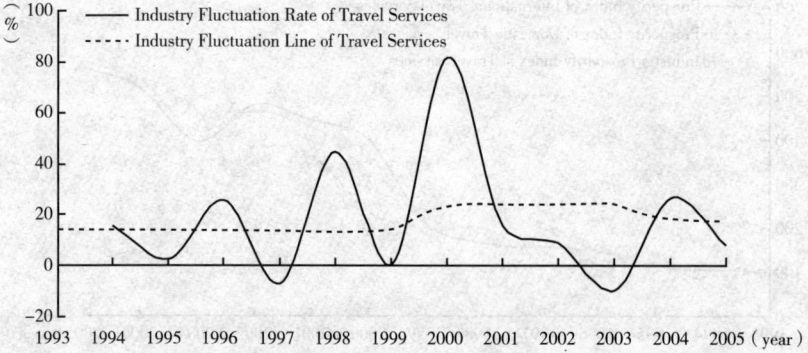

Figure 5   Industry Fluctuation Rates of Travel Services (1993-2005)

Source: *the Yearbook of China Tourism Statistics, the Yearbook of China Tourism Statistics (Supplement), the Yearbook of China Tourism (1994-2006)*

went up and down with a high fluctuation rate, and it is in this new century that the index has shown gentle fluctuations. And with the gentleness of the prosperity line, the travel services industry will gradually make its way for further development and maturation.

4.2   Comparison and verification of research methodology

Using the coefficient method and through the four sub- prosperity indexes of the travel services industry, we can calculate the Industry Prosperity Index of Travel Services. And this index can be compared with the one that is calculated with the coefficient method and through two sub-prospertiy indexes of international

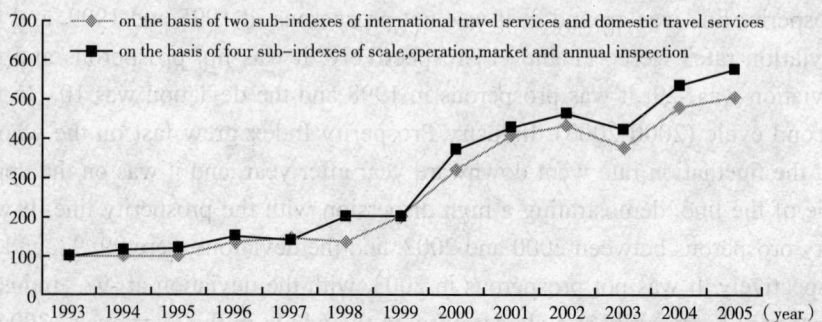

**Figure 6   Comparisons of Industry Prosperity Indexes between
Different Calculation Methods**

Source: *the Yearbook of China Tourism Statistics, the Yearbook of China Tourism Statistics (Supplement), the Yearbook of China Tourism (1994-2006)*

travel services and domestic travel services. The result of comparison is: the former is a little higher than the latter, and the average error is 31. The average annual growth rate of the former is 16.5%, and for the latter, it's 17.09%, both demonstrating the same trend. Taking fators into consideration such as the difference of calculating aperture of Market Prosperity Index, incompleteness of data and the indexes, the two prosperity indexes are highly consistent, which verifies the feasibility of our research methodology.

5. Industry contribution rate analysis

5.1 Internal contribution rate—toursim industry

After 1998, the contribution rate of travel services industry to tourism industry was increasing year after year. It was also true of domestic travel services. However, there was instability in contribution rate of international travel services around 1996, and the rate was on its upward trend after 1998.The average contribution rate of travel services industry to the tourism industry is 0.0918, among this, international travel services 0.0627, more than twice the rate of domestic travel services. It implies that the revenue growth of the travel services industry is faster than that of the whole tourism industry. Compared with the whole tourism industry, the travel industry is prosperous.

5.2 External contribution rate—the third industry and GDP

The contribution rate of travel services industry to the third industry has been growing increasingly since 1997.It is true of the domestic travel services. However, the contribution rate of international travel services was not stable around 1996, but it has grown since 1998. They were all affected by SARS in 2003 and recoverd in 2004,but in 2005,they all fell back to the level before 2000. The average contribution rate of travel services to the third industry is 0.0918,and among these, the international travel services make average contribution by 0.0627,more than twice that of domestic travel services(0.0291). This implies that the revenue of travel services industry grew slightly faster than the overall revenue of the third industry, and the travel services industry was much more prosperous than the third industry, especially between 1997 and 2002.

The contribution rate of the travel services industry to GDP has been growing since 1997; the contribution rate of domestic travel services to GDP has been growing all the time, and it's basically true of the international travel services. They were all affected by SARS in 2003 and recoverd in 2004,but in 2005,they all fell back to the level around 2001. The average contribution rate of the travel services industry to GDP was 0.0041, and among these, the average contribution rate of the international travel services was 0.0041,more than twice that of domestic travel services (0.0013). This implied that the revenue of travel services industry grew slightly faster than that of GDP,and the travel services industry was much more prosperous than the third industry, especially between 1995 and 2002.

With regard to revenue, the growth of travel services industry was slightly faster than that of the tourism industry, and faster than that of the third industry and GDP, which was very especially conspicuous in 1996, 2000 and 2004.

6. Conclusion and prospects

After analyzing the development cycle and prosperity situation of China's travel services industry in many aspects including the overall scale, operational performance, market situation and industry contribution rate, the following characteristics are displayed in the industry operation.

The scale of enterprises within the industry basically fits well with tourism market in China, and there still exists some structural insufficient supply in some segement markets. The market structure went through oligopoly, monopolistic competition, and it's now in the state of complete competition. With regard to industial competitiveness structure,the international travel services are more competitive than domestic travel services,and large ones are more competitive than small ones. This disequilibrium is closely related to the difference in socio-economic development level, touristic infrastructure, touristic accommodation facilities, concepts of operation and management, institutions etc. At present, the vast majority of the subjects of the industry are operating leading products. They are enhancing their specialized operations and conducting tranverse expansion so as to raise their specialization level. Statistics show that the travel services industry is operating with slender profits. However, taking into consideration factors like reasonable tax evasion, incompletion of incentive and restraint mechanism in state-owned enterprises, imperfection of corporate governance and ever-increasing new entrants, there is still much room for the growth of the travel services industry. There are huge gaps between China's travel services and its international equivalents in the elements of core competency including product innovation capability, brand building, network construction.

While China's travel services industry continues to stablize its inbound tourism, there is huge market potential in outbound tourism and domestic tourism. And this sheds some light on market behavior, including the innovation of toursim products of domestic business travel, independent travel, holiday travel, and specialized travel etc and also, China's travel services should attempt for high level of transnational operations while working hard to expand its outbound business and tourism services at the same time.

References

［1］Zarnowitz Victor. *On the dating of business cycles*［J］. Journal of Business, 1963, Vol.36, Issue 2, 179–199.

［2］USGS, *Metal Industry Indicators*, http://minerals.usgs.gov/ minerals /pubs/mii/.

［3］Chen Zhenzhen, Luo Leqin. *Statistics* ［M］.Beijing: Science Press, 2006(in Chinese). 陈珍珍、罗乐勤:《统计学》[M]，北京：科学出版社 , 2006，第 271 页。

An Analysis of the Tourism Value of World Heritages and Study of Their Exploring Model

Liang Xuecheng[1, 2]

(1. Management School, Xi'an Jiaotong University, Xi'an 710049, China; 2. School of Economics and Management, Northwest University, Xi'an 710069, China)

Abstract: This paper, starting from the facts of world heritage tourism in China, points out the internal relationship between the tangible and intangible value of world heritages after analyzing heritage tourism and tourism value. Then based on the hypotheses and their tests, it analyzes and testifies the judgment that exploring the intangible value of world heritages will help to raise their tourism value as well as the satisfactory degree of tourists. Finally, considering the type and feature, it probes into the exploring model and strategy to develop heritage tourism value in China.

Key words: world heritage; heritage tourism; tourism value; exploring model

1. Introduction

Presently, most countries or regions in the world realize the exhibition function of heritage resources through heritage tourism, and heritage tourism has become irreplaceable nameplate products or "gold-lettered signboards" within the

[About author] Mr. Liang Xuecheng(1969—   ), Ph. D. of Management School, Xi'an Jiaotong University, Associate Professor of Northwest University. Research field: Tourism Management.

development of national or regional tourism industry. Xu Songling (2002) points out that opening heritage resources to public via tourism is both a tendency and social responsibility. Guo Zhan (2002) also proposes that the world heritage is the root and base for the development of tourism industry. Up to 2004, China has possessed a total of 30 entries in the World Heritage List, 27 of which are already moderately or intensively exploited, accounting for 90% of total heritage resources. It seems that the practice did partly realize and spread the value of these heritage tourism resources, but it simultaneously creates problems like the over development of the value in heritage resources and the ignorance of heritage protection. What's more, it has a pretty big gap between the apperceive expectation utility of tourists. For tourists, the development of most heritage tourism still stays at the "symbol tourism" stage, lacking deep development and study on the value of heritage tourism, which makes it difficult to offer tourists the natural and cultural connotation of heritage resources. That will inevitably lead to the lack of deep understanding of the value of heritage tourism so that they feel difficult to echo and thus prevent the effective spreading of heritage value. On the other hand, the economic properties of being exterior and public leads to the fact that many tourism products offered by heritage tourism can not be properly adjusted, i.e., can not bring the optimal allocation of resources by the market. Therefore, a simple transformation of world heritage resources to the direction of tourism products, according to the direction of marketing, will bring a bad result that the evaluation of tourism value takes the place of the heritage value, or even the loss of the core values of heritage resources-authenticity and integrity. It may discontinue the inheritance and the use of world heritages, so as to violate the responsibility of protection, preservation and passing down of these heritages as required by *Convention Concerning the Protection of the World Cultural and Natural Heritage* (hereinafter referred to as "*Convention*").

Therefore, to develop tourism value of world heritages needs a full play of the potential role of community policy and explore the value from various aspects. Xu Songling (2003) points out that the value of heritage involves in a variety of areas, including aesthetics, history of thoughts, religion, sociology, history, science and technology, etc. As single heritages, their value type is not entirely consistent with each other. Xie Ninggao (2002) has ever proposed: "the development of world heritages includes two aspects of hard-exploitation (tangible development) and soft-exploitation (intangible development). " But at present, the tourism value of the heritages in most places are just hard-exploitation, which mainly includes the construction of various service facilities and commercial building projects, such as the construction of shops, hotels, restaurants, cables, entertainment cities, etc., whereas the soft-exploitation of their real value is ignored or deviated. Soft-exploitation asks people to explore the scientific, aesthetic, historical and cultural value through research and aesthetic practice, so that it can enhance and deepen the science, education, heritage tourism, including the culture, customs, and other resources. In fact, soft-exploitation of heritage resources is the foundation and source for hard-exploitation, which is also a lifeline to maintain hard-exploitation; hard-exploitation is just a form to realize the derivative value of soft-exploitation. Obviously, the development of tourism value of world heritages does not only need the simple and practical hard-exploitation, but also a more intensive study and exploration of soft-exploitation. Based on the status quo, the author will analyze the world heritage tourism value from two aspects: theoretically analyzing the tourists' satisfaction degree and social welfare resulted from soft-exploitation; discussing developing modes of heritage tourism value, based on the types and features of world heritage resources in China.

2. Literature review

Many scholars home and abroad have been researching the heritage tourism and the exploitation of its value. Since the research perspectives are different, the contents also vary. In explaining the concept of heritage tourism, foreign scholars are more inclined to the attribute of heritage tourism products and the experiential feelings of tourists. For example: Richards (1996a, b) believes that heritage tourism can be defined as a process, or a product. In order to establish a link between the motivation of tourism and heritage tourism, the form of motivation should first be understood. Moscardo (2001:5) holds: "Heritage tourism is an experience arising from the interaction between tourists and resources." Micheal (2002) insists that heritage tourism should not be seen as an arbitrary act, but a social psychological need. Similarly, Poria et al (2001:1048) identified heritage tourism as a phenomenon arising from the feeling of tourists for a special place.

In China, scholars are more inclined to a protective development. For example Wu Bihu (2002) studies the distribution and the needs of Chinese heritage sites and believes that developing the World Heritage Sites into the main local tourism products is a natural choice as well as a result of policy game. Guo Zhan (2002) believes that during the course of heritage tourism, authenticity and integrity must be protected. He is against the separation of the rights in "gate", "protection" and "management" and puts forward an idea of innovative mechanism. Li Rusheng (2002) appeals to establish a law-based protection mechanism to realize the sector management and social supervision; he also suggests to establishing a system of paid use of the World Heritages and getting some money from the charge of the licensed running expenses for protection. Zheng Xiaoxie (2003) proposed that we should "reinforce the protection of heritages and prevent the 'near extinction' problem."

The research of the World Heritage tourism value starts earlier in the foreign countries, and similarly, foreign researches are more linked with the feelings of tourists, i.e. to analyze the exploitation of heritage tourism value from the customer perspective. Richards (1996b: 262) points out that in order to accommodate the preference of human, attention should be paid to the attractiveness of heritages. The uniqueness and polymerization of heritage resources mean that the attractiveness has developed into a special niche of the tourism industry. Teo and Yeoh (1997:193) also hold that as more tourists are attracted to the heritages, its authenticity was pushed to danger so that tourists have to choose destinations according to their expectations instead of the intrinsic value of authenticity. Ashworth and Larkham (1994:16) regard present heritage as a special commodity created purposefully to meet the need of modern consumption.

Heritage tourism exploitation research starts later in China, and most scholars tend to combine the value of heritage tourism with their protection value and tourism value. Wei Xiaoan (2002) views the value of social welfare, and points out that popularization of the welfare with World Heritages must be achieved; the protection and the use of heritages is both the purpose and the mean, and the final result is the promotion of social welfare; the distribution mechanism of public welfare can not be completely formed spontaneously, public management and the implementation of public policies will also help. Xu songling (2003), in his series of papers about Huangshan model evaluation said that the quality and the environment indices should be distinguished and "the pure market-orientation should be altered to the multi-mission directed market operation." He concluded that "dealing with Huangshan according to the entire marketing system is wrong, but a non-profit operation of national park should be advocated."

Generally, present researches on the heritage tourism and its value is more manifested as multi-perspective analysis and interpretation, lacking of further exploration of theoretical analysis of inherent structure and its application model, which is just the focus of this study.

3. The analysis of the world heritage traveling and its value

According to the *Convention*: "World heritages are rare and irreplaceable properties, any corruption or lost of cultural or natural heritage will cause the exhaustion of the world heritages." World heritages, as places of spiritual activities, such as scientific research, education, tour and widening eyesight, is the ideal place of interrelationship between people and nature (Xie Ninggao, 2002). Heritages as a kind of important public resources, the original value lies in its naturalness, which is an important force in maintaining the balance of nature and a history portrayal of human history of civilization. "The core of heritage value lies in its generalized cultural value and knowledge value, which further creates the economic value" and a by-value of tourism development. Xu Songling (2003) believes that the common value of heritages is their generalized cultural value which can be divided into seven aspects. However, from the perspective of compound system of tourism resources, the value of heritages is divided into two aspects: tangible value (explicit) and intangible value (implicit), or four sub-aspects: tourism value, research value, cultural value and environment value (Figure 1):

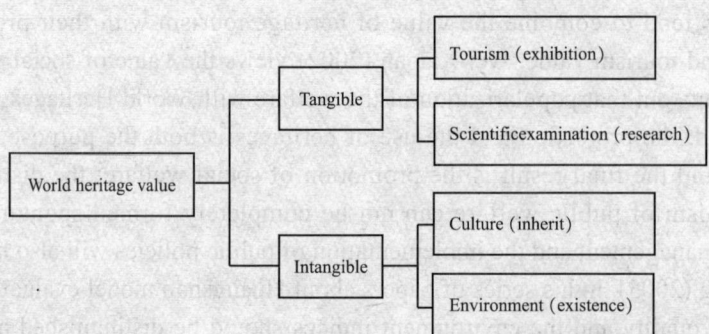

Figure 1 Value constitution of world heritages

Viewed from the value constitution of the world heritage: the tangible value (explicit) value mainly includes the tourism (exhibition) value and the scientific examination (research) value. It is exhaustible, which is regarded as a kind of use value that is reduced gradually through constant exploitation. One typical method in literature is to define use value as economic value relating to the actual use of resources, for example, Smith (1987A) proposes a method by

visiting comfortable scenic spots or observing primitive wild biotic community. The intangible value (implicit), on the other hand, mainly includes culture (inherit) value and environment (existence) value, which is classified as non-use value. They are the sum of value that relies on the "use" part of the total value instead of the pure existence value. John V. Krutilla points out: people possess the unrelated value which has nothing to do with their current resources because it can be used as the reserved option in the future and transfer the natural resource to their successor. Moreover, the total value may be regarded as the payment of a person to protect or maintain current resources. When the total value is more than the use value, the difference is the non-use value, which is also called "the existence value", "intrinsic value" or "retention value". "Joint pledge" divides world heritages into five basic types, and each is characteristic in the value expression. Specifically, there is different inner link and the match relations between the tangible value and the intangible value. The key point of this research will be the different matched pattern between tourism value and cultural value, environment value. Two basic hypothesis are listed as follows: H_{01}—realization and promotion of natural heritage value need the exploitation and transformation of the intangible environment value; H_{02}—realization and promotion of culture heritage value need the exploitation and the transformation of the intangible culture value. Based on H_{01} and H_{02}, hypothesis H_{03} is further proposed: the improvement of exploitation and transformation abilities of the intangible value from heritage resources is helpful in raising tourists' individual benefit standard, which further enhances the social benefit standard.

Firstly, we suppose that tourists may directly enjoy or consume tourism products $Y(y_1, y_2, ..., y_n)$ developed by the inheritance tangible value, which mainly includes tourism service facilities, and its corresponding payment price is $P(p_1, p_2, ..., p_n)$ and tourism product $X (x_1, x_2, ...x_m)$ developed by the inheritance intangible culture (or environment) value, and its corresponding payment price is $R (r_1, r_2, ..., r_m)$, then the inheritance tourism products may be represented as the combination $Q(Y, X)$. Moreover, if we suppose that the maximization of the tourist's individual effectiveness needs to satisfy the constraints $P \cdot Y + R \cdot X = M$, (M is cash income), the tourists' conditioned consumption demand function is $x = x_i(P, M-R \cdot X, X)$; tourists' conditioned indirect utility function is $u = u(P, M-R \cdot X, X)$. Under this utility level, the function may be expressed with the Hickes minimal expenditure: $e^* = M-R \cdot X = e^*(P, X, u)$. If a tourist develops or consumes an inheritance tourism product $q_1(y_1, x_1)$, its price is p_1 and r_1. Suppose x_{min} is the lowest level of product X transformed from the inheritance intangible value, or as it is called the threshold value, then in Fig. 2, the point E is the point of intersection by utility curve u_0 and r the price curve. When $u_0 < u_0 < u_1$, they intersects the value curve r respectively at A, B, C and D, among which A and C enjoys the equal utility level, whereas B and D enjoys the equal utility level. Suppose p* is the highest (or bottleneck) price of inheritance product y_1, and it is the increasing function of product x1, there will be three situations as follow: situation 1 can be explained with formula (1)-If we neglect the development and transformation of inheritance intangible value, there will be no tourism activities at all even if it is open freely; situation 2 can be explained with formula (2), where the consumption of tourists on heritage tourism is positive with this price; situation 3 can be explained with formula (3)-If the payment price is beyond the affordability of tourists, the consumption demand will be zero.

$$\text{When } 0 \leqslant x_1 \leqslant x_{min}, y_1 = y_1(p1, P, x_1, u) = 0 \ (p_1 \geqslant 0) \tag{1}$$

$$\text{When } x_1 \geqslant x_{min}, y_1 = y_1(p_1, P, x_1, u) > 0 (0 < p_1 < p^*) \tag{2}$$

$$\text{When } x_1 \geqslant x_{min}, y_1 = y_1(p_1, P, x_1, u) = 0 \ (p_1 > p^*) \tag{3}$$

$$\text{When } 0 \leqslant x_1 \leqslant x_{min}, y_1 = y_1(p_1, P, x_1, u) = 0,$$

$$\text{One } p_1 \geqslant 0 \tag{1}$$

$$\text{When } x_1 \geqslant x_{min}, y_1 = y_1(p_1, P, x_1, u) > 0,$$

$$\text{One } 0 < p_1 < p^*; \ (2)$$

$$\text{When } x_1 \geqslant x_{min}, y_1 = y_1(p_1, P, x_1, u) = 0;$$

$$\text{One } p_1 > p^*; \tag{3}$$

Obviously, the development of heritage intangible value is helpful for its value realization, which also shows that once the heritage resources is fully market-operated, their prices will quite probably be over the affordability of tourists, resulting in the failure of tourists in enjoying the welfare of heritage tourism, or even make them a luxury product for only a small number of people, which is obviously against the original intention of publicizing the enjoyment of world heritage value. Therefore, the realization of tourism value in heritage value, the attributes and characteristics of heritage resources must be taken into consideration. If the development and transformation of intangible value (cultural or environmental) of the heritage value is neglected, the existence and implementation of the value in heritage resources will definitely be affected. The above demonstration theoretically explains and confirms H_{01} and H_{02}.

Considering promoting the level of tourists' welfare, its measurement can be defined in accord with the changes of X and R. The change of R and the total change of currency income share the same influence. Since it is impossible for individual tourists to adjust the quantity of X so as to meet the optimal

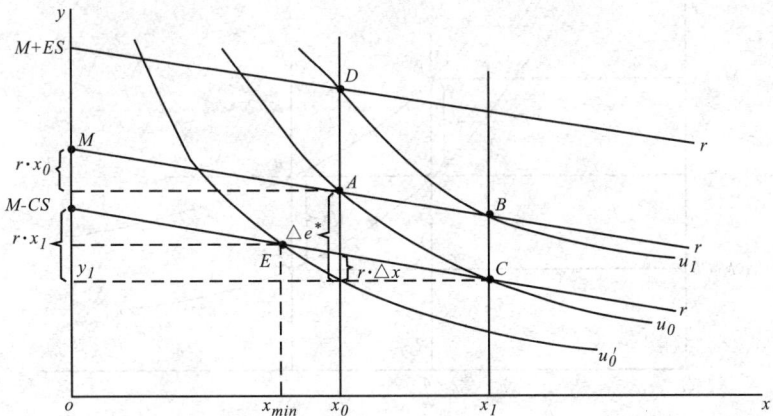

Figure 2   Compensative Surplus with the Increase of x (r=0)

conditions of traditional rate of margin substitution and price ratio equality, the available compensation surplus (CS) and the equal surplus (ES) can be employed to measure relatively the changes of the tourists' welfare. Suppose $x_{min}<x_0<x_1$, with the help of conditioned indirect utility function of tourists, the quantity of CS and ES is the solution of functions (I) and (II).

$$u(p, M-rx_0,x_0)=u(p, M-rx_1-CS,x_1) \qquad (\text{I})$$
$$u(p, M-rx_0+ES,x_0)=u(p, M-rx_1,x_1) \qquad (\text{II})$$

CS means that when tourists have the opportunity to buy new product x1, and the price has changed, tourists need to increase the amount of the expenditure in order to equate the individual welfare with the initial state. In Fig. 3, CS equates with BC, and it can be expressed by formula (III) with expenditure function. ES refers to the amount of increased income when initial price and tourists consumption level x1 are given and one still needs to equate the new price of tourist individual welfare with the price of B. ES equates with AD, and it can be expressed by formula (IV) with expenditure function.

$$CS=e(p, r, x_0, u_0)-e(p, r, x_1, u_0)=M-e(p, r, x_1, u_0) \qquad (\text{III})$$
$$ES=e(p, r, x_0, u_1)-e(p, r, x_0, u_0)=e(p, r, x_0, u_1)-M \qquad (\text{IV})$$

Certainly, if the intangible price is developed and transformed, when the price of a new product X satisfies r=0, the increase of effectiveness obtained from the expenditure of heritage tourism is obvious. That is, the level of effectiveness

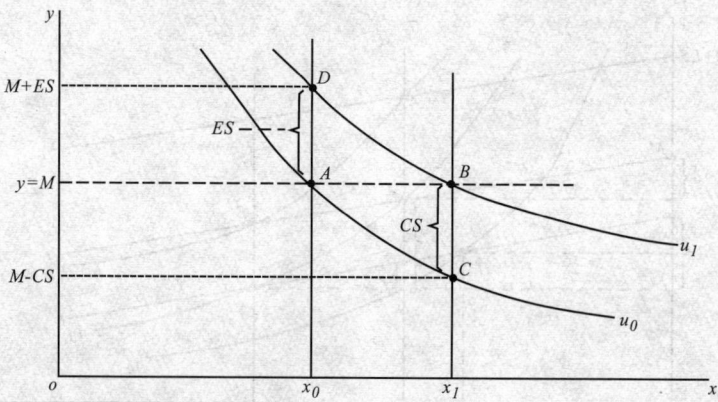

Figure 3   Compensative and Equivalent Surplus with the chang of x (r=0)

promoted from u0 to u1, the level of individual social welfare will be improved (Fig. 3), which will necessarily increase the satisfaction level and the whole level of social benefits. In this way, the original hypothesis H03 has been theoretically explained and tested. This also shows the importance of developing the intangible heritage value. Some foreign researches show that tourists do not travel to the world heritages because of their charm or greatness, but because they are world heritages-there is no feeling of greatness or pride. Therefore, it is quite necessary to improve the intangible value of heritage tourism products in order to help tourists get satisfied with their expectation value.

4. The development model of tourism value of world heritages

Through the tourism value analysis of World heritage, we see that in order to maintain a long life of heritage resources, and let more people truly recognize the value and the greatness of heritage resources, deeper mining of the intangible value and its modality of heritage resources is needed to increase the total value of heritage resources. There are different types of heritage resources, and the developing mode of each type shall not be identical. Here the author will carry out the research from five basic types of the world heritage.

Model 1: Value of natural heritage + Value of developing environment = Realization and promotion of the value of natural heritage tourism

This mode mainly aims at natural heritages. Some scholars have ever thought that to distinguish between natural resources and environmental resources is to gild the lily. Smith (1998, p3) once advocates that "Both natural and environmental resources shall be classified as priced assets" since they can provide various services to human, such as tourism, archaeology, landscape, ecological balance, etc., some of which are interrelated-increasing or decreasing together, but usually the increase of one service will be accompanied by the inevitable decrease of others. When exploiting or utilizing this type of heritage resources, we should minimize the intervention of natural environment; even when we build tourism infrastructures, we should try to avoid damaging the natural environment. The main value of these heritage resources lies in their beautiful natural scenery and harmonious ecological environment, which attracts tourists, like Wulingyuan, Jiuzhaigou, Huanglong, and other natural heritages. Therefore, we should put protection first in the development of natural environment value, and improve scientific means in development. For example, some methods can realize the deep development and utilization of the value of heritage tourism, such as having characteristic or artistic names, taking part in some advertisements concerning environment protection or commonwealth. In addition, for those intangible heritage resources like biographies, religious and cultural stories, or beautiful legends, we can develop some real products with necessary limitations and set them in the transitional areas so that tourists may enjoy them and feel the nature value together with the harmony between human and nature.

Mode 2: Value of cultural heritage + Value of developing culture = Realization and promotion of the value of cultural heritage tourism

The model is mainly directed at the development of cultural heritage resources, including cultural relics, architectures and sites. The number of such heritages is big, accounting for about 2/3 of total heritages in China, and thus is the key of development and utilization. Typical examples are the Forbidden City, Summer Palace, Mausoleum of Qin Shihuang and Terracotta Warriors, Ancient City of Lijiang, Dazu Rock Carving and so on. The value of such heritages lies in its strong cultural characteristic, which is highlighted in the complete true historical description. Since the soul of cultural heritages is the original nature and authenticity, it is non-renewable. In developing intangible value of these heritages, we must focus on the excavation of those stories and events, including the media of audio-video products, such as the creation of films, television programs, and video products. During the course, we should not damage or destroy the original nature. At the same time some heritage-related literature, art, legends and stories, etc., can be created and developed into dynamic and participatory tourism products so that the tourist will better understand the heritages.

Mode 3: Value of both cultural and natural heritages+ Development (cultural value and environmental value) = Realization and promotion of the double value of cultural and natural heritage tourism

This pattern mainly suits the development of dual heritage in cultural and nature values. Although there are not many such heritages as the Yellow Mountain, Tai Mountain, Emei-Leshan Giant Buddha and Wuyi Mountain, etc., they are of high value and reputation. As these resources have a dual attribute, they highly depend on both the culture and the environment value. Therefore, the development of these resources must face certain contradictions, and neglecting any aspect will cause the decrease of their tourism value. Thus, we can never develop these resources without paying attention to their intangible values. However, the fact is some heritage tourism areas in China sacrifice the environment sights for the development of cultural values, or even damage the environment simply for tourism service facilities. Wuliqiao Reservoir of the Yellow Mountain is a typical example. The communities and urbanization of the Yellow Mountain is not conducive to the sustainable development of the heritage resources. Of course, when we focus on the development of environmental values, the cultural values should also be subtly disclosed, which allows tourists to see more than just natural sights. Clearly, the development and management of the dual value of such heritages needs balance and flexibility so as to achieve the harmony, extension and development between the culture value and the environment value.

Mode 4: Cultural sight value + Development of cultural value = Realization and promotion of the value of cultural sight heritage tourism

The mode is about the development of cultural sight heritages. This kind of heritages consist of people's purposeful designs and constructions, organic evolved sights, and the relevant cultural sights, whose obvious potentials lie in intangible cultural values. For example, in the Lu Mountain Area in China, the cultural value is given special prominence; the mode is similar to that of cultural heritage resources. As these heritage resources are somewhat dependent on the natural environment, their development needs our attention to the coordination with the natural environment value.

Mode 5: Value of dictated and immaterial heritage + Help (of activities or stage show) = Realization and promotion of the value of dictated and immaterial heritage

This mode is for immaterial heritages, which are intangible, including cultural activities and oral presentation of special culture, including various expressions of languages, stories, music, dance, games, myths, rituals, customs, folk handicrafts and art. Such heritage resources are few, only two kinds of which in China, namely Kunqu and Chinese paper-cut. The development of such heritage is something special. People appreciate their value through a series of static or dynamic art performance instead of a simple visit. So, their value can be developed to quite valuable tourism activities via festival activities or stage performances, and the especially valuable form is the folk performances related to civil activities, like organizing paper-cut artist to give on-the-spot show. It is also a way to spread our folk culture. The development of such heritage value in China is still in its early stage, and we lack experience in tourism. However, once they are combined with corresponding theme parks or tourism activities, there must be a great consumption market.

In short, one-mode-for-all is not the correct approach. It will be a practical and meaningful process of developing the tourism value of world heritages only when we base our work on the attributes and characteristics of resources.

5. Countermeasure for exploiting tourism value of world heritages

In view of the status quo of exploiting tourism value of world heritages in China, the author proposes the following four countermeasures to guarantee the effective implementation of the development modes:

First, the exploitation should follow the scientific and practical principle and avoid excessive tourism or dislocation development. As a kind of protected resource, world heritages are not purely exploitative resources. If we consider heritage tourism as a golden brand attracting tourists and neglect the relationship of source and the river between the heritage and the tourism, heritage tourism and its value will not be realized or developed smoothly. For instance, the problem of "too many houses in the cultural heritage spots" and "artificialization, commercialization, and urbanization" of natural heritage site has become great threats to the development of these heritage resources. Therefore, a systematic view is needed in developing heritages tourism value, which needs to pay attention to both tangible value and intangible value. This will be helpful to the realization and promotion of the heritage tourism value, and to the protection and long-term development of heritage resources.

Second, scientific plans are needed in developing heritage resources. No matter which mode is chosen, a set of scientific planning is a must, including the division of functional areas and design of activities, and scientific planning measures shall be adopted. At heritage sites, protection measures, division of functional areas, opening zones, and service establishments are all essential to the planning. Lacking of scientific method to define environmental capacity of heritage resources, many heritage sites still have difficulties in determining the reasonable number of their tourists. However, there must be a division of heritage sites according to a certain standard, so as to alter the current situation that tourists "visit the mountain and live at the foot of mountain" into "visit inside while live outside", creating a friendly eco-environment circle. In addition, there seems to be lack of anything unique or shock-evoking in heritage sites. Tourists usually feel that they need something like a prelude before visiting or an epilogue, an echo, or sublimation in the end. The missing of these important links is bound to inhibit the enhancement of heritage tourism value. Therefore, in order to increase the degree of tourists' satisfaction and improve the social welfare from heritage tourism, the intangible value of heritage resources must be developed and planned scientifically.

Third, protective measures and mechanisms should be enhanced. The development of heritage value needs protection, which is not just the responsibility of governments, but also of the whole society. In fact the funds for heritage protection is far from inadequate, and to make it worse, with the further development of economy, there will be more new artifacts found, and the work of unearthing, preserving and researching needs great input of labor, financial and material support. Since these supporting forces are limited, a variety of forces or forms from the community shall be encouraged to participate in developing heritage tourism value. At the same time, we should ensure that the basic attributes of cultural heritages are the premise to safeguard its original value attributes. Therefore, China is in great need of establishing a mechanism for the protection of heritage development and avoiding the state that the protection and the developing were isolated. What's more, the subversive buildings in natural heritage sites and the excessive tourists in cultural heritage sites shall not exist any longer.

Fourth, laws and regulations shall be employed in heritage protection. Committees of experts, organizations, and management system of reformation should be established to protect world heritages through the legislation management, for example, *The Law of National Heritage Protection* shall be constituted, and National Heritage Administration Bureau shall be established as soon as possible. As cultural and natural heritages are related to many disciplines, it is highly scientific with great social benefit. We should adopt ideas from experts of various disciplines before appropriate assessments and evaluation. Thus it is essential to organize a committee of experts on heritage protection. As heritage resources show great diversity in their value and it is unique in the economic and social attributes, there is a widespread controversy in heritage resources management system. Therefore, it will be safer and more appropriate to improve corresponding organizations based on pilot reforming.

6. Conclusions and further studies

World Heritage has multiple values. The analysis and development of heritage tourism value embody the scientific exhibition and enjoyment functions. The research on heritage development shall not be limited within material and tangible ones, but more attention should be paid to the hidden or intangible value so that the heritage value can be realized and upgraded. Certainly, to develop the intangible value of heritage resources further is not only beneficial to advance heritage tourism, but also directs us to a new way to realize heritage value, which brings us many benefits. To be specific, it will benefit sustainable development, offer managers an innovative approach of developing, and satisfy tourists' expectant demands. However, in the specific process of developing and transforming heritage values, we need to break through many traditional restraints, including the system, institution, culture, psychology, and so on. In addition, there bound to exist problems concerning the coordination of market environment, government behaviors, consumer preferences and other aspects during the management of world heritage resources, which may be a further study field.

Generally, only when we make more detailed analysis and research on this problem can we make the value of heritage resources widely recognized and spread, and further guarantee and encourage more people or groups to participate in the protection, development, and enjoyment of these heritage resources.

References

[1] Ashworth G, J P Larkham, (eds). Building A New Europe: Tourism, Culture and Identity

in the New Europe［M］, *Routledge, London*. 1994

［2］Michael E. Antiques and Tourism in Australia［J］. *Tourism Management*, 2002, (23): 117–125.

［3］Moscardo G. Chapter1: Cultural and Heritage Tourism: The Great Debates. In, B Faulkner G Moscardo, E Law, (eds). *Tourism in the 21st Century: Lessons from Experience* London: Continuum. 2001. 3–16.

［4］Poria Y R, Butler Airey D. Clarifying Heritage Tourism［J］. *Annals of Tourism Research*, 2001, (28):1047–1049.

［5］Richards G. Cultural Tourism in Europe［M］. *Wallingford: CABI publishing*. 1996.

［6］Richards G. Production and Consumption of European Cultural Tourism［J］. *Annals of Tourism Research*, 1996, 32: 261–283.

［7］Teo P, Yeoh B.Remarking Local Heritage for Tourism［J］. *Annals of Tourism Research*, 1997, 24: 102–213.

［8］Freeman A Myrick. The Measurement Of Environmental and Resource Values–Theory and Methods［M］. Zeng Xiangang Trans. Beijing: China Remin University Press. 2002. A. 迈里克 · 弗里曼：《环境与资源价值评估——理论与方法》［M］，曾贤刚译，北京，中国人民大学出版社，2002。

［9］Li Rusheng, Guo Zhan, Xie Ninggao, Wei Xiaoan, Zhang Baoyun. Study on China's Tourism Development［J］.Tourism Tribune,2002, (6): 5–9(in Chinese). 李如生、郭旃、谢凝高、魏小安、张宝赟：《中国旅游发展笔谈》［J］，《旅游学刊》2002 年第 6 期，第 5~9 页。

［10］Liang Xuecheng, Hao Suo. An Analysis of the Value in Constructing a Compound Tourism Resource System［J］. Tourism Tribune, 2004, (1): 61–66(in Chinese). 梁学成、郝索：《对旅游复合资源系统的价值分析》［J］，《旅游学刊》2004 年第 1 期，第 61~66 页。

［11］Liang Xuecheng. Theoretical Analysis and Demonstrative Study on Shaanxi's Application for World Heritage Tourism Attraction Exploitation［Z］2002.6. 梁学成：《陕西省申报世界遗产与旅游资源开发的理论分析与实证研究》［Z］，2002.6。

［12］Tao Wei. Research on the Sustainable Tourist Development of "World Heritage" in China［J］. *Tourism Tribune*, 2000, (5): 35–41(in Chinese). 陶伟：《中国 "世界遗产" 的可持续旅游发展研究》［J］，《旅游学刊》2000 年第 5 期，第 35~41 页。

［13］Wang Xingbin. The Reform of Managerial Modes of China's Natural and Culture Horitage［J］. *Tourism Tribune*, 2002, (5): 15–21 (in Chinese). 王兴斌：《中国自然文化遗产管理模式的改革》［J］，《旅游学刊》2002 年第 5 期，第 15~21 页。

［14］Wu Bihu, Li Mimi, Huang Guoping. A study on relationship of conservation and tourism demand of world heritage sites in China［J］. *Geographical Research*, 2002, (5): 617–625(in Chinese). 吴必虎、李咪咪、黄国平：《中国世界遗产地保护与旅游需求关系》［J］，《地理研究》2002 年第 5 期，第 617~625 页。

［15］Xie Ninggao. On Cableway's Threat to World Heritage［J］. *Tourism Tribune*, 2000, (6): 57–60(in Chinese). 谢凝高：《索道对世界遗产的威胁》［J］，《旅游学刊》2000 年第 6 期，第 57~60 页。

［16］Xu Songling.On the Management of World Heritage in China—the Evaluation and

Renewal of Huangshan Mode ［J］. *Tourism Tribune*, 2002, (6): 10–18(in Chinese). 徐嵩龄:《中国的世界遗产管理之路——黄山模式评价及其更新 (上)》［J］,《旅游学刊》2002 年第 6 期, 第 10~18 页。

［17］Xu Songling.On the Management of World Heritage in China—the Evaluation and Renewal of Huangshan Mode ［J］. *Tourism Tribune*, 2003, (1): 44–50(in Chinese). 徐嵩龄:《中国的世界遗产管理之路——黄山模式评价及其更新 (中)》［J］,《旅游学刊》2003 年第 1 期, 第 44~50 页。

［18］Xu Songling.On the Management of World Heritage in China––the Evaluation and Renewal of Huangshan Mode ［J］. *Tourism Tribune*, 2003, (2): 52–58(in Chinese). 徐嵩龄:《中国的世界遗产管理之路——黄山模式评价及其更新 (下)》［J］,《旅游学刊》2003 年第 2 期, 第 52~58 页。

［19］Xu Songling. On the Selection of the Operational System of China's Heritage Tourism Industry—On Separating Four Kinds of Powers and Balancing Them ［J］. *Tourism Tribune*, 2003, (4): 30–37(in Chinese). 徐嵩龄:《中国遗产旅游业的经营制度选择——兼评 "四权分离与制衡" 主张》［J］,《旅游学刊》2003 年第 4 期, 第 30~37 页。

［20］Xu Songling. Reform of Management Institution of Chinese Culture and Natural Heritages ［J］. *Management World*, 2003, (6): 63–72 (in Chinese). 徐嵩龄:《中国文化与自然遗产的管理体制改革》［J］,《管理世界》2003 年第 6 期, 第 63~72 页。

［21］Zhang Chengyu, Xie Ninggao. The Principles of Authenticity and Integrity and the Conservation of the World Heritage ［J］. *Journal of Peking University(Humanities and Social Sciences)*, 2003,(3):62–68(in Chinese). 张成渝、谢凝高:《"真实性和完整性" 原则与世界遗产保护》［J］,2003 年第 3 期《北京大学学报》(哲学社会科学版), 第 62~68 页。

［22］Zheng Xiaoxie. Strengthen World Natural and Cultural Relics Protection and Prevent Them from "Imminent Danger" ［J］. *Urban Studies*. 2003, (2): 50–54(in Chinese). 郑孝燮:《加强我国的世界遗产保护与防止 "濒危" 的问题》［J］,《城市发展研究》2003 年第 2 期, 第 50~54 页。

［23］Chinese World Heritage Website. http://www.cnwh.org(in Chinese). 中国遗产网 . http://www.cnwh.org

New Perspective of Improving the Quality of Tour Guide Services

—— Concurrently an Overview of Guide Service Management and Research over the Past 20 Years

Wang Jing, Ma Yaofeng

(College of Tourism and Environment, Shaanxi Normal University, Xi'an 710062, China)

Abstract: Based on the analysis of the authoritative data of investigations by the National Tourism Bureau, the paper points out that the quality of guide services is the shortest "plank" in tour services system in China. The researches done in recent years into the problems of guide services haven't come to the roots of them both in practice and in theory. This paper suggests a new perspective of tourist experiences, which is different form the perspective of guide supervision to improve guide services. It further puts forward some suggestions to reform the management of guide services based on the analysis of the problems related to tourist experiences, quality of guide services, and system of tour guide management.

Key words: guide; management system; service quality; tourist experience

Introduction

The domesticv tourism sample survey made by the joint work force of Urban

[About authors] Ms. Wang Jing (1973-), A doctor candidate majoring in tourism management in Shaanxi Normal University, E-mail: wjj7306@163.com; Mr. Ma Yaofeng (1949-), a professor majoring in tourism planning and GIS in Shaanxi Normal University.

Socio-economic Survey Organization and Rural Socio-economic Survey
Organization under the jurisdiction of the National Tourism Bureau and National
Statistics Bureau surveyed 10229 urban residents and 4967 rural residents
respectively. The survey of/on the quality of domestic tour services involves
seven elements/items: accommodation, transportation, food, entertainment,
shopping, tourist attractions, and guide services. The fact that the guide services
ranks last in the Table of Evaluation on Tour Services by Urban Residents and
Rural Residents indicates that guide services has become the least satisfactory
service item domestic tourists (see Table 1).

Table 1 Evaluation on Different Tour Services Experienied by Urban Residents and Rural Residents (%)

Evaluation Items	Urban Residents		Rural Residents	
	Very good	Good	Very Good	Good
Overall Impression	8.0	50.4	5.84	49.37
Accommodation	6.0	21.7	4.17	20.74
Transportation	11.9	49.5	14.05	48.96
Food	6.4	35.9	4.19	28.85
Entertainment	5.4	23.9	2.72	12.48
Shopping	5.0	24.9	7.59	39.16
Tourist Attractions	11.8	41.2	4.43	19.67
Guide Services	2.7	12.0	1.23	6.80

Source: Data from the Sample Survey of Chinese Domestic Tourists 2004[2].

In the survey on respondents who have used tour services (see Table 2), the
percentage of "Poor Impression" on guide services is largely/basically higher
than those on other tour service items. That is to say, for those respondents who
have used tour services being surveyed, guide services has the worst reputation.
Based on the comprehensive analysis of Table 1 and Table 2, it's obvious that
guide services lags behind and hinders the other domestic tour service items.

Table 2 The Percentage of "Poor Impression" on Tour Service Items by Urban Residents and Rural Residents as Services Users (%)

	Overall Impression	Accommodation	Tran-Sportation	Food	Enter-Tainment	Shopping	Tourist Attractions	Guide Services
Urban Residents	0.73	1.01	1.05	1.28	1.34	1.42	0.53	2.27
Rural Residents	0.23	0.79	0.50	1.04	0.69	0.66	0.96	0.99

Source: Data from the Sample Survey of Chinese Domestic Tourists 2004[2].

Table 3 indicates that the percentage of "Never Use" of guide services is the highest among all the seven service items, accounting for 73.6% of those urban respondents and 85.86% of those rural respondents respectively. The low percentage of using guide services, on one hand, attributes to the fact that guide services doesn't belong to basic tourist consumption items and therefore isn't a necessary part of tourism supply. On the other hand, it is due to the fact that tourists choose not to use guide services because of their poor impression of guide services and the low social reputation of guide services.

Table 3　The Percentage of "Never User "of Different Tour Services over the Total Number of Urban Residents and Rural Residents Surveyed (%)

	Overall Impression	Accommo-dation	Tran-Sportation	Food	Enter-Tainment	Shopping	Tourist Attractions	Guide Services
Urban Residents	4.6	50.6	4.6	14.1	47.9	36.5	24.9	73.6
Rural Residents	3.3	54.19	3.19	25.81	70.81	20.74	66.76	85.86

Source: Data from the Sample Survey of Chinese Domestic Tourists 2004[2].

As one of the service items provided for tourists, guide service takes up a core position among all the tour services and always plays a leading role. Guide services links travel agencies, tourists and other tour operations as a strong bond. The quality of guide services is symbolic of the quality of tour services in a tourist destination.[1] The authoritative survey made by the National Tourism Bureau sounded the alarm for the guide services in China from all angles. The poor reputation, low satisfaction, low percentage of the use of guide services distinctly exposed themselves in the growth of domestic tourism in China. The quality of guide services is the shortest "plank" in tour services system in China. Just as the shortest plank decides the volume of a barrel, so the quality of guide services harms the overall quality of tour services. Therefore, the quality of guide services is worth academic attention and research and those problems related to it must be promptly solved.

1. A summary of tour guide management and related researches in China

The research into guide services in China falls into two perspectives. One involves the managerial practice of tour guides that reflects the history of the industry of tour guiding. This is the current background in which this paper holds a discussion on the problems of guide services. The other perspective makes a research into the specialized theoretical research on tour guides, shedding light on the hot theoretic points on guide services in different growth stages esp. on the identification, reasons and practical guidance of the problems emerging in guide services.

1.1 A review of the management of tour guide as a profession in China

Since the founding of the People's Republic of China in 1949, tour guiding as a job has been initiated. In those days, tour guides was part of the foreign reception staff with a limited number of 200 to 300. Tour guides at the time was originally called "interpreter guide" who worked with China International Travel Service or China Travel Service and provided services to groups visiting China from friendly countries.

Since 1980, with China's reform and opening to the outside world and the development of economy, the booming domestic tourism and international tourism have been asking for ever increasing tour services, thus the tourism supply has been expanding rapidly. By the end of 2004, the number of tour guides who passed Tour Guide Qualification Examination has reached 220000. China began to make laws and regulations on the management of tour guide profession in 1980, which gradually developed into a corresponding management system.

1.1.1 The major laws and regulations laid down on tour guide management in China (in time order)
Temporary Provisions of Guides Management, March 1,1988 (terminated in 1999)

Guides Profession Grade Standard (Temp), 1994

Guides Services Quality (National Standard) GB/T15971-1995, 1995

Requirements of Domestic Tourism Services Quality of Chinese Travel Agency, 1997

Regulations of Guides Management, Oct. 1, 1999

Measures for the Implementation of Regulations of Guides Management, Jan. 1, 2002

Measures for the Management of Guide Qualification Certificate, 2002

Measures for the Management of Guides Profession Rating (Temp); Measures for the Implementation of Regulations of Guides Management (Revised), July 3, 2005

1.1.2 The system of tour guide management in China

Since 1990s, China has developed a sound and healthy management systems of the tour guide qualification authentication, grading and examination management, IC card pointing management, yearend evaluation on tour guide performance, centering around Regulations of Guides Management and Measures for the Implementation of Regulations of Guides Management.

The laws and regulations are being enforced by different managerial functions, basically forming an effective daily management of tour guides. The IC card pointing management on tour guide beginning 2002 is viewed as a revolutionary leap in the management of tour guides because it exerts a dynamic supervision and administration on tour guides.

1.2 A summary of researches into guide services in China

By looking up *Chinese National Knowledge Infrastructure*, CNKI database on December 2, 2006 by typing in "tour guide", 685 papers were founded. After deleting 18 irrelative papers or repeated papers, altogether 640 academic papers on tour guides or tour guiding were available which were published during the period 1986 to 2006. These papers reflect on the status quo of the academic researches into tour guide services in the past 21 years.

1.2.1 Conclusions can be made based on the analysis of Table 4: The academic researches into tour guide services were primarily amassed in the period of 13 years since 1994, during which the number of published academic works account for 97% of the total in the past 21 years. Since 1994, the domestic tourism in China has embarked on rapid growth, meanwhile, guide services has been gradually supervised and managed by law. However, from 1994 to the present, the market demand for tour guides has become greater while guide services is in comparatively short supply, thus many related problems arising. In addition, the newly laid down guide services management system is not complete and needs further adjustment and exploration. In line with this, researches into guide services theory began to be active. Owing to the incompleteness or imperfection of tourism theory, the theoretic researches into guide services closely follow the hot points or pressing problems in tourism practice. As a result, the research topics in theoretic researches actually make known those pressing problems in guide services that call for urgent attention and solution.

1.2.2 The research topics in Table 4 can be further classified into 7 categories:

(1) Introduction to guide's job (including job description, feature, function,

Table 4 A Review of 640 Specialized Papers on Tour Guiding over the Past 21 Years

Year	Papers (Quantity)	Theme
1986	1	Tour Guide Data Base
1987	2	Guide Skills, Investigation on Regional Guide Team
1988	1	Tour Guide Psychological Quality
1989	3	Guide Speech/Text, Service Psychology, Advanced Deeds/Anecdotes
1990	1	Introduction to Guide's Job
1991	0	-
1992	1	Guide Language
1993	5	Spot Guide, Language, Cross-cultural Communication, Specialized Needs
1994	13	Hierarchical Management, Presentation, Tour Map, Profession Socialization, Teaching, Specialized Need, Advanced Anecdotes/Deeds
1995	8	Overseas Experience, Aesthetic Quality, Advanced Anecdotes/Deeds, Spot Guide, Culture Guiding
1996	18	Specialized Teaching, Image Creation, Styles, Spot Guide, Competence and Quality, Service Standard, Emotional Quality, Guiding Art
1997	26	Tour Map, Advanced Anecdotes/Deeds, Teaching, Spot Guide, Language, Realm, Specialized Needs, Robot Guide
1998	18	Teaching, E-guide, Language, Promotion, Guide History, Tour Map, Tourist Attraction Guiding, Team Investigation, External Experience, Complaint
1999	24	Teaching, Spot Guide, health and Sanitation, Computer Guide, Discussion on Management Rules, Language, Advanced Anecdotes/Deeds, Language of Presentation, Function and Influence
2000	44	Overseas Experience, Network Guide, Language, Teaching, Image and Skills, Advanced Anecdotes/Deeds, Income, Quality, Supervision and Administration, Discussion on Management Rules, Modern Technology, Spot Guide, Reform of System, Training, Tour Map, Complaint, Market Order
2001	28	Function, Advanced Anecdotes/Deeds, Language Translation, Guide Speech/Text, Professional Quality, Virtual Guide, Laws and Regulations, Examination, Complaint, Morality, Art, Overseas Guide, Language, Aesthetics, Health and Sanitation, Tour Map, E-guide

Continued Table 4

Year	Papers (Quantity)	Theme
2002	58	Function, Pay, Spot Guide, Advanced Anecdotes/Deeds, Rebate, Shopping Complaint, E-guide, Training, Language, Cross-cultural Communication, Regional Team Construction, Aesthetics, Overseas Experience, Specialized Needs, Work Ethic, Teaching, Ecology and Environmental Protection, Examination, Image, Management, Knowledge Structure, Quality Requirement, Pointing Management
2003	51	Quality, Profession Mechanism, Profession Classification, Pay System, Teaching, Training, Overseas Experience, Guide history, Service Creation, E-guide and Software, Qualification Examination, Charisma, Language, Law and Regulation, Training, Team Investigation and Construction, Morality, IC Card Management, Service Quality, Law Breaking and Complaint
2004	87	Quality, Electronic Information Technology, Guide Speech/Text, Specialized Needs, Tour Map, Rebate, Complaint, Teaching, Tiredness of Work, Communication Deviation, Guide Psychology, Language, Guide Management, Communication, Characteristic, Service Quality, Tip, Examination, Presentation Art, Aesthetics, Pay, Knowledge Structure, IC Card Reform, Market Management, Training, Ecological Culture
2005	101	E-guide System, Spot Guide, Teaching, Advanced Anecdote/Deeds, Heath and Sanitation, Tiredness of Profession, humanism Education, Guide Services Company, Career Positioning, Pay System, Presentation, Reform of Management System, Body Language, Mobile Phone Guide, Ecological Requirement, Marketing Awareness, Income Survey, Language, Geographical Guiding System, Overseas Experience and Comparison, Team Construction, Creative Work, Service Quality, Grading and Promotion, Discipline Violation, Education and Training, Personality and Character, Qualification Authentication, Cross-cultural Communication, Aesthetics
2006	150	E-guide, Guide Information System, Aesthetics Accomplishment, Guide Verbal Language, Excellent Guide, Tour Guide Major Teaching and Education, Service Quality, Tourist Satisfaction, Market Analysis, Guide Social Needs, Regional Investigation findings, Pay-Rebate-Tip-Grey Income, Survival Situation, Work Satisfaction, Human Resources Tapping, Presentation and Quality, Work Ethic, Profession Right Protection and Benefit Expression, Development and Operation of Guide Services Company, Regulation Violation and Compensation, Guide Examination etc.

influence, marketing function etc.)

(2) Policy and law (guide management regulation, service standard, qualification examination, IC card pointing system)

(3) Guide's personal quality and skill (guide's language, interpersonal communication, cross-cultural communication, knowledge structure, psychology quality, professional skill, work ethic, creative service, advanced anecdotes and deeds. Guide's language includes translation, presentation, guide speech while psychology quality includes competence, character, emotion, and professional skill includes image, style, realm and aesthetics.)

(4) Specialized needs and creative services (garden, ecology, environmental protection, popular science, specialized cultural tour)

(5) Management (guide socialization, team construction, qualification examination, guide services company, grading and promotion, professional positioning, market order and complaint, human resources management, work state and satisfaction, tourist satisfaction, pay etc.)

(6) Tool and modern technology (tour map, robot technology, network information technology)

(7) Education, teaching, training and others

1.2.3 It's discovered that the most popular topics concentrate on two topic categories: tour guide's personal quality and skill, and management based on a comparison of the appearance frequency of all the research topics in Table 4. The researches into guide's personal quality and skill have been popular throughout the years, while those hot research topics relative to management usually appear accompanying the enforcement of new laws and regulations on management of tour guides.

Theoretically speaking, the quality of guide services in tourism practice should have improved in line with the gradual perfection of the tour guide management systems and the academic achievements concerning guide services so far reached. Unfortunately, what has been expected has been seen. Although accurate comparative data are not available, it is positive that the service level or standard of tour guides hasn't increased in pace with the growth of tourism economy. On the contrary, it has been declining. This is a conclusion reached based on the facts that guide services ranks last in the Table of Evaluation on Tour Services by Urban Residents and Rural Residents and that society holds negative or unfavorable opinions of tour guide as a profession. Therefore, a complete, in-depth, thorough analysis of guide services should be made to find out those crucial problems affecting its quality, so that the key reform point which is expected to effectively improve the overall quality of tour services will be found out and focused attention on. Only when the bottleneck resulting from guide services has been broken through, can tourism services as a whole take a huge step forward.

2. Why these research topics are put forward

2.1 The major problems concerning guide services in China at present fall into three categories:

2.1.1 Tour guide's personal quality
(1) Lack of knowledge reflected in tour presentation

Little tour presentation or "dumb guide", too many ridiculous fairy stories, lack of knowledge of the cultural value, historic value, and scientific value of tourist attractions, groundless presentation, repeated and over simple presentation embodied themselves notoriously in guide's presentation when it comes to conducting a tour.

(2) Poor professional quality reflected in guide services

The accidents like missing a tour group, and meeting it at a transportation depot late which undergo frequent complaints from tourists indicate that guides are not open to changes, and lack of the ability to plan well ahead, not having a strong sense of professional responsibility.

Change of itinerary or adding more tourist shopping items without permission and things like that are undergoing frequent complaints reflect in a degree that some tour guides don't have good personal quality and work ethic.

Some tour guides respond negatively to tourist complaints or deal with them in an inappropriate way. This results in a harsher conflict between tourists and tourist product supplier, even greater loss on the both sides. Cases like this suggest that tour guides can't live up to the qualified professional skills and work ethics they are supposed to have had.

(3) Poor comprehensive quality of guides reflected in the communication process between guides and tourists

Some tour guides are not able to communicate effectively with tourists. What they focus on is just economic gains and benefits during the process of interacting with tourists. Hence, tourists generally hold such negative opinions of guides as slippery, over smart, dishonest. What was described above proves that the comprehensive quality of guides is poor and that guides don't have a good mastery of professional skills.

2.1.2 Problems concerning tourism management amassed in the process of guide services

(1) Disputes relative to "group fee" between an organizing agency and a receiving agency

Disputes relative to "group fee" between an organizing agency and a receiving agency are the issue being most frequently complaint over. When such problems arise, guides will be the first ones involved. If a guide deals with the dispute inappropriately, he/she will be subjected to tourist complaints, thus becoming a black goat in the "group fee" dispute of the two travel agencies.

(2) Contract dispute between tourists and travel agency

The National Tourism Bureau and those tourism administrations at provincial level have drafted sample tourist contract and stipulated that a contract must be signed between tourists and travel agency they use. Unfortunately, there are still some agencies that will not sign a contract with their customers. Some agencies even set traps in contract clauses to deceive tourists. During a tour, if any problem does arise, the folks on the front line – tour guides will be the first ones to be blamed for they represent a travel agency. Sometimes, even if a lawful tour contract has been signed, conflicts still show up because it is not effectively fulfilled. In this circumstance, guides are still inevitably subject to tourist complaints as one of the persons who are carrying out the contract.

(3) The lowered reception standard resulting from the short tourism supply in peak season

In peak season, due to the short tourism supply, tourists may not enjoy good tourist experience and satisfactory tourist comfort. This is beyond the control or capacity of guides. However, guides are still the ones to be blamed by unsatisfied tourists for they are representatives of tourism suppliers.

Guides take up a special position in tourist industry, carrying out the commitments on visitors on behalf of travel agencies while safeguarding tourist rights by supervising tourism suppliers. Therefore, tour guides is a coordinator trying to satisfy both tourists and tour suppliers. They subjected themselves to travel agency's employment and tourist complaint. Under the condition that both travel agency and tourists give their right and power a full play, the problems relative to guide services largely derive from a guide's poor professional quality. In tourism practice, the case is that neither travel agency can choose and hire guides freely nor can tourists exercise their right or power to complain fully. Consequently, the problems concerning guide services in a large part attribute to the ill functioning system of guide management.

2.1.3   Problems related to guide management
(1) The positioning of tour guide as a profession

Liu Xintian's findings (2005) [3] show that according to the current profession classifying system in China, the guide's job defies any professional category. It is classified into the subcategory of personnel providing services involved in tourist industry and sightseeing section. As guide's job doesn't belong to any professional classification, so tour guides have no professional titles. The Tour Guide Grading System laid down by the National Tourism Bureau has not been effectively enforced over the years. Furthermore, the titles for tour guides didn't gain an entry into the national professional titles alignment, not exactly corresponding to the junior, intermediate, senior titles in professional titles authentication. The titles for tour guides only exist in tourism enterprises, which may not bring corresponding pay or bonus to them.

The current situation of the positioning of tour guide profession directly lead to the fact that the social status of guide is low and that tourism talents with high quality have no intention of taking up tourism as a career. Meanwhile, the threshold for Tour Guide Qualification Examination is relatively low (anyone with senior high school education is entitled to take the exam). As a result, a large number of examinees irrationally took the exam so that an exam rush has been created. After they got the certificate, most guides attach more importance to how to make more money, instead of how to do their job well. Additionally, the prospect of being a guide is not as promising. Those people involved don't take a long-term view. They just intend to earn more and quicker money while they are young. Given a chance, they will hop to other jobs. Therefore, few guides make efforts to further sharpen their professional skills and to make researches into tourism culture.

(2) Under who's jurisdiction

What section should guides subordinate to not only concerns the identity of tour guides, but also involves a series of issues like guide supervision, daily administration, training, awarding and punishment, promotion and benefits and things like that.

In line with the management of tourism market in China, a guide is forbidden to organize and receive tour groups working on his own. If so doing, he will be punished. Therefore, at present, tour guides in China are not free lancers in the real sense. They are under the jurisdiction of travel agency or tour guide service company.

The guides who are under the jurisdiction of travel agency can be further classified into two categories: one is full-time guide working with travel agency who can get pay and may do other jobs for the company besides conducting tours. They can enjoy a steady salary and bonus and receive regular training and subject themselves to daily administration from the company. Their competence is strong, work experience rich, and quality high. Most guides with senior professional titles fall into this category.

The other is "attached guide". They have to deposit their certificates with the travel agency for the opportunity to receive tour groups in the name of agency. They are actually part time guides conducting tours out of interests, for internships, or to earn extra money. Owing to lack of regular training and supervision, guides in the second category comparatively don't have very good professional skills and experience. Most of them are guides with junior titles. They are a supplementary force when travel agency is short of hands in peak seasons.

As a matter of fact, when it comes to tourism high season, guides with high quality and rich experiences are urgently needed while those "attached guides" are not eligible for the tough job involved.

Tour guide service company has been recently set up in some hot tourist cities, which serves as a middle man providing services to those who have passed the Test of Guide Qualification but can't become an "attached guide" for various reasons. Consequently, these guides can receive regular training, even punishment, award and promotion corresponding to their work performance if they submit a certain amount of money to it. On the other hand, travel agencies may also turn to tour guide service companies for hiring guides when they need additional hands. Tour guide service company charges fee for providing intermediary services to both travel agencies and guides. In reality, many problems arise concerning tour guide service companies. For example, they may only charge guides management fee but will not offer them chances to receive groups, not to say provide training for them. Another problem involved is that travel agencies are inclined to look for guides on their own instead of turning to tour guide service companies for help, in that guides recommended by them are not eligible or competent.

(3) Guide's pay

The findings of the Invention Group of Tourism Department of Nan Kai University [4] indicate: the income structure of guides is unreasonable for guides don't get any pay from travel agency or only receive a minimal pay. Conversely, they are supposed to pay a so called "head fee" to travel agency according to the current mechanism of pay and salary for tour guides in its narrow sense. The income of guides mainly comes from rebates or kickbacks, which however is regarded as unlawful or illegitimate. The irrational income structure in fact encourages guides to bring tourists to more shopping arcades, which directly leads to problems relative to guide services in China's tourist industry.

The current guide's pay mechanism directly results in the deviation of guide's value orientation, thus hindering the improvement of guide team's quality, steadiness, work ethic, etc.

(4) Punishment, award and promotion

Punishment and award is one of the means of supervising or administrating guides. The current IC card pointing system stipulates that points will be deducted if a guide breaks relative laws or regulations. When ten points altogether have been deducted, the guide's certificate will be suspended. IC card policy plays a role in the field of guide management or supervision. But it doesn't have a wide coverage and strict restriction. It can't exert a restrictive influence on the majority of guides, esp. those working in off-the-way scenic areas.

2.2 The analysis of the causes of problems concerning guide service quality

From the perspective of tourism management, the supervision over guide services quality is carried out through two channels at present: One are the law executors from management bodies. They enforce supervision over guides through IC card pointing system. The other is tourist complaint and its handling. The problems reflected in these two channels are the key ones which need immediate solution. However, they cannot shed light on guide service quality. The supervision primarily aims at those unlawful behaviors of guides, or at the so called "black guides" in tourism market, not at the quality of guide services quality. As for tourist complaint, it is only an option tourists will turn to when involved in disputes or clashes with huge economic loss. The purpose of the complaint is to gain economic compensation, which can't reflect whether the guide services are good or not. Therefore, that a tourist hasn't filed a complaint doesn't mean that he is satisfied with the guide services he used.

As a result, to analyze the quality of guide services deeply and fully, not only the perspective of tourism management should be adopted, but also the perspective of tourist experience. It can be safely said that changes must be made when it comes to the perspective of analysis and research so that the quality of guide services will be improved radically, i.e., the perspective of tourism management should be completely shifted to the one of tourist experience. This paper suggests that the roots of the problems concerning guide services should be dug out from the perspective of tourist experience, that the guide services quality should be improved by reforming the system of tourism management. This is the basic thinking by which the problems of guide services in China will be effectively solved.

Accordingly, a figure has been worked out to analyze the roots of the problems concerning guide services quality, aiming to solve the above discussed problems reflected in the three aspects of guide services. The paper puts forward a new, creative perspective of researching into guide management, i.e., the hierarchical analytical method based on tourist experience, service quality, and management system.

A one-way or two-way arrow indicates cause and effect or mutual impact while a hollow arrow indicates tracing to its source or getting to the root of the matter.

3. Conclusions and suggestions

The above Figure 1 shows that the problems concerning guide services quality mainly result from the system of guide management, esp. from *Test of Guide Qualification*, the system of professional titles and grading, the system of daily supervision and training, and the reform and perfection of pay and salary. Based on the results of an in-depth survey on tourists, on the part of tourist experience,

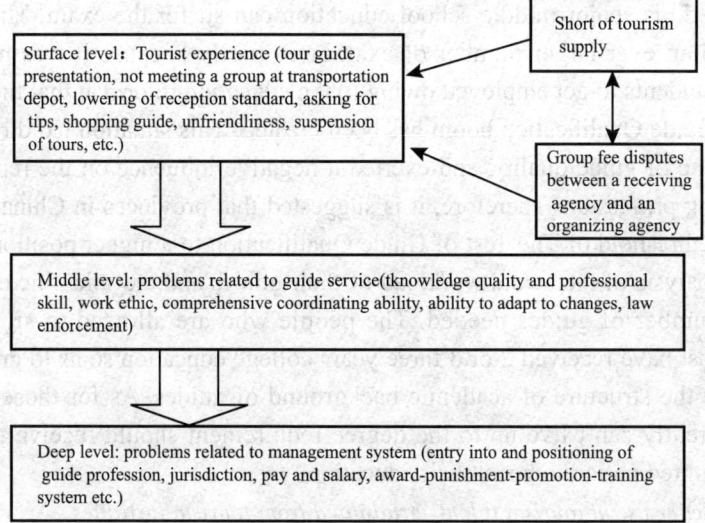

**Figure 1   The Analysis of the Problems Concerning Guide Services Quality
from the Perspective of Tourist Experience**

the management of guides should focus on the continuous improvement of guides' personal quality and enhancement of their work ethic. As for whether a guide works part time or full time, tourists consider it as not so important. Some tourists even welcome those guides from different backgrounds or walks of life who are supposed to offer more highlights in their services and tour presentations. Therefore, to improve the quality of guide services, the improvement of tourist experience should be focused. From this perspective, the management of guides is expected to go through the following reforms or adjustments.

3.1 Choosing talents for tour guide force

The choosing of guides should be based on personal quality and knowledge to lay a solid foundation for the overall quality of the whole guide force.

3.1.1 The threshold of test of guide qualification be raised

The Test of Guide Qualification begun in 1989 intending to boost the quality of guide force which aims at those guides working with travel agency who are required to bear a guide qualification certificate to receive groups. Since 1990s, due the rapid increase of international and domestic tourists, tourism market has been falling short of guides. To solve this problem, the exam was open to anyone with an associate degree beginning 1996. Since 2000, the domestic holiday tour has begun to prevail and needs more guides. In line with this, anyone who has completed his senior middle school education can sit for the exam. The exam attracted an ever-larger number of examinees partly because it was hard for college students to get employed owing to the education reform at that time. The Test of Guide Qualification boom has been created. This situation led directly to the decline of guide quality, and exerted a negative influence on the reputation of guiding profession. Therefore, it is suggested that provinces in China should raise the threshold of The Test of Guide Qualification to a higher position based on the analysis of the actual workload of their guides and the scientific estimate of the number of guides needed. The people who are allowed to sit for the exam must have received 2 or 3 three years college education so as to gradually optimize the structure of academic background of guides. As for those guides who currently can't live up to the degree requirement should receive training within limited period to boost their education.

3.1.2 Better use of tourism talents graduated from tourism institutes

By the end of ninth five-year plan, there are 1195 tourism institutes in China with 330000 undergraduates altogether. The talents trained by tourism institutes characterize themselves by all-round knowledge and skill relative to tourism, and better personal quality so that they should be fully employed by travel industry. It's a common practice that students studying in teachers colleges and universities are expected to apply for and get teachers' certificate while still in college. [5] Students majoring in tourism and related fields should also follow suit to get guides' certificate for their future career.

3.1.3 Employing both professional and non-professional guide

The experts who study and understand culture, history, tourism, architecture, biology, and science and technology fall into the category of expert guide. Thanks to their rich knowledge and broadened vision, their contact with visitors will be impressive and cherished on the part of tourists. As for those with either special experience of or contact with a scenic spot or tourist attraction can be employed as non-professional guide. On the part of visitors, the contact with non-professional guides can help deepen their understanding of the culture of a tourist destination, consequently better their tourist experience.

3.2 Management of learning-oriented tour guides

The internal management of tour guides focuses on service skills, occupational morality and the motivation of life-long learning. The major measures adopted may include the tour guide grading system, the extensive excellence rewarding and multiple-access learning rewarding system, and the internal rewarding system through accumulated scoring etc.

3.2.1 Stress on the training of service skills and the formation of occupational morality

When tour guides are certified with qualifications after passing the certificate tests, the administrative agencies of all levels should pay attention to the training of tour guiding service skills and formation of occupational morality. Specifically, example or model tour guides are to be set up, outstanding tour guides with excellent service skills and magnificent occupational morality are to be honored, and workshops for service skills and morality are to be regularly held. Tour guides who have done well in service skills innovation and set great morality example should be honored on an extensive basis, as more example tour guides will facilitate to develop a self-discipline and progressive atmosphere.

3.2. 2 Motivation of life-long learning through various rewarding systems

A variety of rewards should be established concerning guiding knowledge, service skills and morality traits, and the review mechanism may consist of the mutual review between tour guides, the review by visitors, tour operators and the tour guide administrative agencies. To bring the rewarding system into full effectiveness, the rewards should be diversified to cover about 50% of all tour guides. And the rewards are better combinations of material benefits and spiritual encouragement, particularly the welfare benefits such as increase of salary, increase of the housing accumulation fund and endowment insurance, more opportunities for further education, same and stable salary for the outstanding liberal-occupation guides, etc. which helps motivate tour guides to improve their service and stabilize the outstanding force of tour guides.

3.2.3 Career plan for tour guides

People both inside and outside the tour guide force popularly regard tour guide as an occupation for the young. This has been a myopic view that may lead to a negative atmosphere for learning and service improving, little initiative to be example and model guides, and serious employment mobility. Career planning will help improve the integral service level of the tour guide force through formation of a life-long occupation desire and promotion of career satisfaction. To help tour guides develop a long career objective, what is the most imperative currently is to establish the tour guide grading system, by which a rank hierarchy for tour guides with the primary title, the intermediate title, the senior title and the supreme title will be set up.

3.3 Rationalization of the pay system to improve the occupational reputation of tour guide

3.3.1 Salary and welfare to excellent tour guides by travel agency

There was once a strong voice to pay tour guides regular salary, in order to restrain them from seeking kickbacks and tips. This was only to be found hard to put into practice. For example, the regulation of Hainan Province that the fixed monthly salary paid to tour guides should be no lower than RMB300 turned out simply to be an executive embarrassment. As a matter of fact, it is really hard for the travel agency to pay regular salary to a liberal occupation tour guide. However, it would not be an embarrassment for travel agencies to pay outstanding tour guides who provide regular guiding services for them fixed salary and welfare fringes like full-time tour guides. What's more, this is an effective rewarding mechanism for the travel agencies to employ and motivate excellent tour guides.

3.3.2 Awards to tour guides by the administrative bodies

The administrative organizations like tourism administrations, tour guide service companies, travel agencies, etc. should encourage tour guides with awards on an extensive scale. These awards, as a regular income to tour guides, are not only an important part of their salary, but also an effective method to encourage tour guides to improve their services.

3.3.3 Kickbacks be prohibited, commissions managed in an open manner while tips be gradually advocated.

Kickbacks or rebates, commissions and tips are three different concepts. Kickbacks, coming from the extra-markup, are sort of unfair income to tour guides and harmful to the travel industry in the long run. Therefore, the China's Regulations on the Administration of Tour Guide clearly stipulate that tour guides not be allowed to receive any kickbacks. Commissions, a kind of intermediary charge for travel agencies paid by related business in the tourist product chain, are actually the main source of income to the travel agencies when their business reaches the maturity stage. Therefore, commissions are mainly for the travel agencies, not for tour guides. Tips are a small sum of money given as an award by consumers to people providing quality services. Tour guides may take tips so far as their services win tourists' favorable comments and tourists are willing to give. What's more, tips encourage tour guides for better services and giving tips has been popular in the west. Nevertheless, Chinese consumers are still not used to giving tips to people providing services. Therefore, tips should be gradually advocated in China as the internationalization awareness increases.

3.4 The service quality supervision of tour guide be enhanced with more attention to tourist experience and better use of the IC card

In 2002, China adopted IC card policy in tour guide management, which has been an effective method in terms of tour service supervision and the routine management. However, this method has not been seriously implemented due to a number of reasons. Firstly, due to the equipment or personnel reasons, the governmental inspection over the implementation of the IC card policy was at a rather low level, about 20% in average through out the country. Some provinces and municipalities even never adopted it. Secondly, the inspection has been mainly on the violations of the related regulations, rather than the tour service quality, i.e., the qualified tour guides in the eyes of the governmental bodies were those that did not break the regulations, which is far different from the qualified tour guides in terms of visitor satisfaction. Thirdly, the penalty degree has been rather light and the scope very limited. Take Shaanxi for example, which has been comparatively doing a better job in this aspect, the number of those who got penalty tickets accounted for only 2.4% of all inspected in 2005. The reasons for these problems were partially the loose inspection of the governmental bodies, more importantly, the full functional potential of the IC card has not been brought into play, the detailed supervision items have not been made specific in terms of the tourist experiences and satisfaction, and attention was not fully attached to the service quality assessment of tourists. These have to be reconsidered to improve of the IC card policy.

Thank my good friend Ms. Zhang Yan for her translation this paper into English.

References

[1] Tao Hanjun. Conspectus of Guides Services [M]. Beijing: Chinese Tour Publishing Company, 2003. (in Chinese) 陶汉军:《导游服务学概论》[M], 中国旅游出版社, 2003.

[2] Chinese National Tourism Bureau. *Data from the Sample Survey of Chinese Domestic Tourists 2004* [M]. (in Chinese) 中国国家旅游局:《中国国内旅游抽样调查资料 2004》[M], 北京:中国旅游出版社, 2005, 第76~162页。

[3] Liu Xintian. Key to Guide's Management Practice of Our Country——the Occupation Define and Salary System [J]. *Journal of Yibin University*, 2005, (3).(in Chinese) 刘辛田:《我国导游管理实践中的关键——职业定位和薪金制度》[J],《宜宾学院学报》Journal of Yibin University, 2005, (3), 第48~50页。

[4] The Invention Group of Tourism Department in NanKai University. A Discussion about Establishing Rational Occupation Mechanism for Tour Guides [J]. *Tourism Tribune*, 2003, (6)(in Chinese). 南开大学国际商学院旅游学系本科生科研创新小组:《对建立合理的导游人员职业机制问题的探讨》[J],《旅游学刊》2003年第6期, 第71~76页。

[5] Wu Yanji. A Study of Another Approach to Confirm the Qualification of Tour Guide [J]. *Journal of Zhangzhou Teachers College* (Philosophy & Social Sciences), 2005, (3). (in Chinese) 伍延基:《导游人员资格认定的另一种途径》[J],《漳州师范学院学报》(哲社版) 2005年第3期, 第25~28页。

Development of the Domestic Tourism Education Research by the Statistical Analysis on the Issue in Chinese Version

Zhao Peng, Yu Jifeng

(Institute of Tourism, Beijing Union University, Beijing 100101, China)

Abstract: The source of this article refers to the China Journals Full–text Database and relevant index articles published during 1979 to Dec. 2006. It analyzes such information about these articles as time, district, issuer, author, content and journals and makes a perspective study on the development, problem and trend of domestic tourism education in recent 30 years.

Key words: Chinese issue; tourism education research ; article analysis

Domestic tourism education research is developed along with the development of the tourism education. In recent years, the tourism education scale in China has been expanded rapidly. From 2001 to 2005, the number of the tourism colleges has increased 184 and the tourism institutes cover all the provinces, autonomous regions and municipal cities in China. By 2005, there are 1336 tourism colleges with over 560000 [1] students. For many years, the tourism education has introduced a lot of talent to Chinese tourism industry and promoted the tourism

[Fund] This article is one of the stage achievement of the key subject entitled Chinese Tourism High Education Development and Study of the Eleventh Five Year Plan in China Association of Higher Education (No. 06AII0010076).

[About authors] Mr. Zhao Peng (1954-), president of Institute of Tourism, Beijing Union University, researcher focusing on tourism higher education study; Ms. Yu Jifeng (1971-), master, assistant researcher in Institute of Tourism, Beijing Union University.

education research.

Through article index, we find that in the study of the domestic tourism education research, there have been no articles indexed and analyzed from China Journals Full-text Database. To optimize the effect on the information source and to investigate the situation of domestic tourism education research by quantitative analysis, this article will study, arrange and analyze the articles published in Chinese Issue related to tourism education in the recent 30 years to make a perspective study on its development, problem and trend.

1. Method and result

This article will refer to Chinese Journals Full-text Database as information statistics source and select 3 key words including tourism education, tourism major and tourism talent by using literature measure. Meanwhile, it has searched all the articles, titles, authors, key words, summary and full context published from 1979 to 2006 and sorted it for analysis and statistics.

1.1 Publication time and numbers

Chinese tourism education began in 1970's. According to the index, we find that there are 1537 domestic articles covering such key words as tourism education, tourism major or tourism talent in 30 years including 412 core issues. From the volume of the thesis, the domestic education study was seldom found before 1994. Its first jumping development was in 1994 and the total volume of the thesis reached 36, equal to the total volume of the previous 5 years (1989-1993). Since then, the volume of thesis on domestic tourism education research has been increasing gradually; in 2003, the second jumping development appeared with its volume of thesis exceeding 150; in 2004, its volume was over 200 (please see Table 1 for data). In terms of the statistics of the academic trend in the China Journal Network, the academic and users' attention paid to domestic tourism education has increased.

1.2 Issuing units

The high-output issuing units [2] refer to those units producing no less than 4 theses. According to the index, there are 83 issuing units producing over 4 theses in China. From Table 2 listing the units publishing over 5 theses (including 5),

---

① The number of the tourism institute in this article includes the institutes of tourism and other institutes having tourism major.

Table 1   Volume of Thesis in Past Years

Year	Vol.	Year	Vol.
1980	1	1996	47
1982	1	1997	52
1985	3	1998	73
1986	3	1999	105
1987	5	2000	85
1988	11	2001	93
1989	7	2002	95
1990	6	2003	154
1991	6	2004	230
1992	5	2005	238
1993	2	2006	252
1994	36	Total	1537
1995	27		

Note: the statistical data is up to Dec. 20, 2006. The same below

Table 2   Unites Issuing Above 5 Theses

No.	Issuing Units	Vol.
1	Beijing Institute of Tourism	48
2	Guilin Institute of Tourism	32
3	Jinan University	18
4	Hubei University	17
5	China National Tourism Administration	15
6	Shenyang University	14
7	Liaoning Normal University	12
8	Tourism College of Zhejiang	12
9	Chongqign Institute of Technology	12
10	Dongbei University of Finance and Economics	12
11	Beijing International Studies University	11
12	Guangzhou University	10
13	Shenzhen Polytechnic College	10
14	Sichuan Normal University	10
15	Chongqing Education College	10
16	Guangdong University of Business Studies	9
17	Jianghan University	9
18	Jiangxi Science & Technology Normal University	9

Continued Table 2

No.	Issuing Units	Vol.
19	Nankai University	9
20	Northsouthern Normal University	9
21	Zhejiang Normal University	9
22	Chongqing Institute of Technology	8
23	Yiwu Industrial & Commercial College	8
24	Hunan Normal University	8
25	Jishou University	8
26	Huaqiao University	7
27	Shanghai Institute of Tourism	7
28	Urumqi Vocational University	7
29	Wuhan Institute of Technology	7
30	Zhongshan University	6
31	Chinese Academy of Social Sciences	6
32	Xi'an Jiaotong University	6
33	Wuxi Institute of Technology	6
34	Shanghai Normal University	6
35	Shann'xi Normal University	6
36	Inner Mongolia Normal University	6
37	Luoyang Normal University	6
38	Lujiang Vocational University	6
39	Anhui University	6
40	Chengdu University	5
41	Hengyang Normal Institute	5
42	Hunan Institute of Technology	5
43	Hunan University of Commerce	5
44	Huangshan College	5
45	Nanyang Normal University	5
46	Shaoguan University	5
47	Shunde Institute of Technology	5
48	Taiyuan Normal University	5
49	Northwestern Normal University	5
50	Southwestern Jiaotong University	5
51	Xinjiang University	5
52	Yunnan University of Finance and Economics	5
53	Yunnan TV University	5
54	Zhengzhou University	5
55	China Forestry College	5

it can be found that Beijing Institute of Tourism and Guilin Institute of Tourism, as the early established tourism institutes, are in the leading position in the field of tourism education research with more theses. There are 28 unites producing 4 theses including East China Normal University, Capital Normal University, Northwestern University, Xiamen University, Lanzhou University, etc.

1.3 Issuing regions

Similar to the unbalanced regional development of domestic tourism education, the regional difference in tourism education research is great. There are 11 regions with their issuance over 50 and the numbers of the tourism institutes in those 11 regions are in the top of the nation. There are 7 regions with their issuance less than 10 and the numbers of the tourism institutes in Tibet, Qinghai, Ningxia and Gansu are less than 10.

Table 3   Statistics on the Regional Issuance Volume

Region	Vol.	Region	Vol.	Region	Vol.
Beijing	130	Chongqing	42	Jiangxi	13
Hunan	92	Henan	41	Inner Mongolia	11
Zhejiang	86	Guizhou	40	Gansu	9
Liaoning	79	Shandong	33	Jilin	7
Guangdong	77	Shanghai	29	Tianjin	5
Sichuan	73	Shanxi	27	Hainan	5
Jiangsu	70	Helongjia	26	Qinghai	4
Hubei	69	Xinjiang	26	Tibet	3
Guangxi	55	Anhui	24	Ningxia	2
Fujian	53	Shann'xi	23		
Yunnan	51	Hebei	23		

1.4   Journals

In China Journal Full-Text Database, there are 54 journals issuing over 5 theses including 11 core journals. The Tourism Tribune organized by Beijing Institute of Tourism is the only National Chinese Core Journal in the field of domestic tourism study. Meanwhile, it is the journal publishing the most articles related to tourism education research. Since 1986, the Tourism Journal have published 8 issues of specially titled Tourism Talent and Education Material in consecutive years. The special publication attracts many domestic teachers and scholars engaged in tourism education and becomes the most important platform for the tourism education research. Also it improves the progress of the national tourism education research.

Table 4   Journal Publishing Above (Including) 5 Articles

No.	Journal	Vol.
1	Tourism Tribune *	319
2	Journal of Guilin Institute of Tourism	79
3	Vocational & Technical Education Forum	35
4	Tourism Science	27
5	Journal of Beijing International Studies University	26
6	Chinese Vocational and Technical Education *	25
7	Education and Vocation *	22
8	Social Scientist	13
9	Market Modernization *	10
10	Vocational and Technical Education	10
11	Journal of Hunan Economic Management College	10
12	Journal of Chongqing Institute of Technology	10
13	Liaojing Education Research*	9
14	Middle Vocational Education	9
15	Vocational Education Research	9
16	Communication of Vocational Education	8
17	Journal of Urumqi Vocational University	8
18	Inquiry into Economic Issues *	7
19	Journal of Guangxi Education College	7
20	Development	7
21	Journal of Chengdu Education College	6
22	Higher Education Exploration *	6
23	Enterprise Economy *	6
24	Journal of Nantong Vocational University	6
25	Journal of Jiangxi Science & Technology Normal University	6
26	Journal of Guizhou University (Social Science Edition)	6
27	Journal of Guangxi Normal University (Philosophy and Social Science Edition)	6
28	Heilongjiang Researches on Higher Education	5
29	Yunnan Researches on Higher Education	5
30	The Border Economy and Culture	5
31	Journal of Guizhou Ethnic Institute (Philosophy and Social Science Edition)	5
32	China Three Gorges Construction	5
33	Chinese Forestry Education	5
34	Journal of Zhejiang Institute of Tourism	5
35	Journal of Yuxi Normal College	5
36	Journal of Xinjiang Vocational University	5

Continued Table 4

No.	Journal	Vol.
37	Journal of Wuhan University of Science and Engineering	5
38	Journal of Taian Teachers College	5
39	Journal of Suzhou Vocational University	5
40	Journal of Sichuan University (Social Science Edition)	5
41	Research *	5
42	Journal of Chongqing Education College	5
43	Journal of Hunan First Normal College	5
44	Hotel Modernization	5
45	Journal of Huaqiao University	5
46	Journal of Jinlin Vocational University	5
47	Science & Technology Economy Market	5
48	Science Information (Academic Version)	5
49	Journal of Leshan Normal College	5
50	Journal of Liaoning Senior Vocational School	5
51	Journal of Luoyang Normal College	5
52	Journal of Mingjiang Vocational University	5
53	Journal of Nanchang Vocational & Technical Teachers College	5
54	Journal of Ningbo University (Educational Science Edition)	5

Note: * is core journal.

1.5 Issue content

In all the issues, the such top 4 researches are tourism education and development research, regional tourism educational research, talent cultivation research and professional education research. Among these, the subject of tourism education and development research is the problem and solution of the tourism education, in which most of the articles analyzing the current situation, features, external environment, reform suggestion and strategic thinking. The content of regional tourism education research mainly centers on the development between the tourism education and regional industry coordination and the problem, opportunity and development strategy of the regional tourism education. The talent cultivation research is to mainly study the talent cultivation pattern, current situation and strategy along with the problems occurred in the quality cultivation and special tourism talent (i.e. tour guide, sports tourism, conference, etc). The professional education research focuses on the pattern, method, problem, content, way and reform of the professional education. Column "Others" in Table 5 refers to information, conference, document, notice, book, advertisement, etc.

Table 5   Statistics of Issues Categories

No.	Category	Vol.	No.	Category	Vol.
1	Development Study	239	10	College and Feature Introduction	27
2	Regional Tourism Education	228	11	Educational Material Study	26
3	Talent Cultivation	219	12	Education Research	26
4	Professional Education	217	13	Student	25
5	Reform on Course and Education Content	78	14	Faculty Research	21
6	Practical Education	57	15	Quality and Assessment Study	12
7	Professional Construction	56	16	Postgraduate Education	11
8	Foreign Tourism Education	56	17	Others	194
9	Education and Talent Market	45			

1.6   Fund Thesis

It is found by the statistics that the thesis on fund project has been published since 1997 and the volume of the theses has increased year by year. According to the data in 2005 and 2006, the relevant results in fund project was increased remarkably, which revealed that the tourism education research had been paid attention and financed (see Table 6).

Table 6   Statistics on Fund Project Results

Year	State Level (vol.)	Provincial Level (vol.)	Municipal Level (vol.)	College Level (vol.)
2006	4	18	2	5
2005	6	24	1	8
2004	0	14	0	7
2003	0	5	0	4
2002	0	0	0	3
2001	0	1	0	1
2000	0	3	0	0
1997	0	1	0	0

1.7   Authors

There are 1418 authors involving in 1537 theses. Table 7 indicates the number of authors publishing over (including) 2 theses, which can help us know the

Table 7 Statistics on Authors' Theses Publication

Vol. of Thesis	2	3	4	above (including)5	Remarks
Number of Authors	179	56	14	19	98 theses published by unknown authors

units and individuals having strong capacity in this field, especially the authors publishing over 5 theses shall be the main study power in the said field.

1.8 Number of Reference

68 out of 1537 theses is referred and please see Table 8. There are 30 theses introduced by above 3 times, which influences the trend of academic development. There are 3 theses introduced by above 10 times, in which the Feasibility and Necessity of Increasing Sport Tourism Major in Social Sports Major (Han Lu'an, etc) and Journal of Tianjin University of Sport (Issue 1, 1999) was referred by 21 time; Tourism Study in American University (II) – Tourism and Related Major (Wu Bihu, etc. Tourism Tribune, Issue 3, 2001) was referred by 16 times; The Current Situation of Conference Education in International Institutes of Tourism and Difference in China (Jin Hui, Tourism Tribune, Issue 1, 2003) was referred by 13 times.

Table 8 Statistics of Reference Frequency

Frequency	1	2	3	4	5	6	7	8	above 10 times
Vol.	26	12	14	4	6	1	1	1	3

2. Problem analysis

The academic articles are the media, which can rapidly reflect the academic development trend. The key words selected in this article are limited in tourism education, tourism major and tourism talent. In case of any tiny change in the key words, the volume and content of the indexed article will be changed accordingly. However, the statistical information source in this article is from China Journals Full-text Database to ensure the enough samples. Therefore, the missing and cross will be existed, but it has sufficient reference value in the overview, current situation, disadvantage and trend in the analysis of the domestic tourism education research.

2.1 The rapid development of domestic tourism education research is becoming the relatively independent study field

The volume of articles in tourism education research has increased year by year since 1994 and it keeps high increasing ratio. From the content, it involves in such fields as tourism education development, regional tourism education, talent cultivation, professional education, professional construction, reform on the curriculum and education content, teaching1 material, practical education, college establishment, talent and market, faculty and postgraduate education. In the recent 3 years, the articles focusing the foreign study have increased, which enable the domestic tourism education put on the dual channels of domestic observation and overseas attraction. The tourism education has received more and more concerns and has become a relatively independent study field.

2.2 The tourism education research comes into being together with tourism education, which reflecting the stage features of the tourism education

The first thesis related to tourism education research in China was in 1980, which was published by Zhang Danzi, entitled Suggestions on Compiling Tourism Education Context on Issue 1 of Social Science, 1980. From its issuance time, it can be found that people start study it when the tourism education is created.

Horizontally, Table 5 in this article reflects the issue content of the tourism education research. Vertically, in the few theses published before 1997, most of them studied on development (especially the problem and solution); from 2003, the talent cultivation research became the hot topic, especially during 2004 to 2006, it enjoyed high frequency of publication, which was related to the rapid expansion of Chinese tourism education scale. The regional tourism education research is always in leading position. Since 2004, with the high talent demand in the regional tourism economic development, such research has had faster development. The volume of the theses during 2004 to 2006 was over 100.

2.3 The tourism education research has an academic group of certain scale, but lack of strong power

Compared with the volume of the issues, there is huge author group in the field of tourism education research. They are management, teachers, researcher in research institution, experts and leaders in governmental authorities and some scholars.

By statistics, over 85% authors are at the beginning of the tourism education research. There are two reasons: one is that it was not untill 2004, the results in fund project began to increase (occupied 10% of the volume), which indicates shortage of the expense and the group was unstable. The low cost concept study was popular. The other is most of authors were teachers or researchers in tourism major, but education study was not their major. Therefore, few of them could insist on the middle & long term study. Compared with other education study fields, it lacks of the strong power.

2.4 Wide study field, but less study type, study focus and the depth of the theory to be strengthened

By reviewing the issues, it covers many results. But in terms of the way of study, its study type mainly focuses study description. Other types including comparison study, experiment study and historical study were few. In terms of the objective of the study, it was mainly focusing on the development study but lack of fundamental study, assessment study and prediction study. Although there are many articles related to the current situation of the tourism education, problem and solution and how to develop the tourism education and economy, objectively, there are few hot topics to be discussed in the tourism education field.

From anther point of view, the expansion of the tourism education research didn't bring the real prosperity. Many articles are from the job description and the various personal thoughts and feelings with less depth in theory. The expansion and depth in the study process failed to be synchronized. The above-mentioned study went against the sustainability and depth of the study, which was repeated in low level in recent years; meanwhile, the disperse of the study power leads to the slowly improvement of the whole study. It is hard to produce great and significant study result.

2.5 Study blank and future trend

The domestic tourism education has been continuously developed, renovated and deepened. The content of the education is enriched and its view is expanded. By the statistics analysis, the following subjects are not only the blank or weak fields, but also the content to be further strengthened and concerned in the future: (1) fundamental theory study of the tourism education; (2) demonstration study between the tourism education and economic development relationship; (3) the study of tourism education modernization; (4) quality assessment study on the tourism education; (5) faculty development study in tourism major; (6) foreign tourism education analysis and reference study (the overseas introduction began to increase from 2004, but it's still not enough especially it lacks of the localized study on the foreign matured tourism education mode).

As we know, the education study can improve the development of the tourism education in each stage, but the tourism education shall be further developed by the modern way of education study to ensure its features of renovation and creation. Therefore, we shall absorb the new fruits in the education study field by its study method, which is a future trend for tourism education research.

3. Conclusion: tourism education research–subject dialog and value integration

The volume of articles related to the tourism education research will be increasing, which objectively reflects that the domestic tourism education is in the development. From the issuance in recent years, it brings passion and interest than ever before and many experts join in the study and discussion. We can hear different voices from different subjects (government, industry, college and research institutions) giving their opinions on the tourism education from different point of views.

Besides the detailed topics, the discussion made by many subjects on tourism education is the ideal discussion. It is the ideal exchange and dialogue of tourism education between subjects and is the free expression of their value. The exchange and dialogue are the perpetual process; therefore, it is the process for dialogues with many subjects and many opinions, which becomes the important factor to influence the development of the tourism education.

The articles related to tourism education are created from the free dialogues. The increasing in the volume indicates the attraction of the tourism education. But the common interest brings different opinions due to different subjects. Its unique demand and special judgment will bring various ideal expectations for the tourism education. However, the tourism education exists and is developing. From the way of thinking in Analytic Philosophy, the recognition of education is to project the real education phenomenon into the educational language system. The one who can control the correlation correctly can have the right of explanation. Therefore, the real significance to the tourism education research is the sustainable pursuance and exploration to the development of the tourism education.

There is certain distance from academic discussion to practical exploration until the practice. The tourism education only has 30-year history in China, its content shall be enriched from the subject dialogues and educational practices.

For the development of tourism education, we really need a power to integrate the social value and concept. So far, various institutes of tourism of various levels have been set up. The value integrity is more important. The guidance of curriculum position, major arrangement, talent cultivation target and specification, curriculum systems and the regular supervision to the educational activities in tourism major shall be of great importance. Therefore, we shall analyze and use the existed study results and optimally integrate the social values to guide and improve the tourism education, and make its study root in the theory study and reform practice to have strong life power.

References

[1] China National Tourist Administration. Basic information of national institute of tourism in 2005. (In Chinese) 国家旅游局:《2005 年全国旅游院校基本情况》[EB/OL], http: //www. cnta. com.

[2] Wu Shulian. A perspective on evaluating the current situation of the Chinese higher education by the articles related to the evaluation on the Chinese higher education [J] . *Chinese higher education evaluation*, 1998, (3). (In Chinese) 武书连:《从中国高等教育评估发文情况透视中国高教评估现状》[J],《中国高等教育评估》1998 年第 3 期。

Several Discriminations about Rural Tourism, Agro–tourism and Folklore Tourism

Liu Deqian

(Institute of Tourism, Beijing Union University, Beijing 100101, China)

Abstract: Rural tourism, agro–tourism (or farm tourism) and folklore tourism are increasingly catching the attention of both the tourism industry and academic circle in China. However, the one–sided understanding of them in research papers, conference speeches and tourism planning reports are not only rather general but also spreading abroad. The author of this article hold that, if tourism industry and academic circle could focus more widely on the practices of the industry and explore further based on former research, it will probably take more positive roles on theories and practices of rural tourism, Agro–tourism and folklore tourism. Concerning the development of these three forms of tourism in China and relevant theoretic issues, there are several points of this article which are personal explanations and discriminations the author made according to his nearly 20 years' observations.

Key words: rural tourism; agro–tourism; folklore tourism; theory and practice

1. Discrimination about rural tourism

1.1 Starting from the content

If you open Web Pages of "Beijing Rural Tourism Website, 2005" (http//ly.

[About author] Mr. Liu Deqian, professor of Institute of Tourism, Beijing Union University, research fellow and deputy-director of Tourism Research Center, Chinese Academy of Social Science.

binw. gov. cn) sponsored by Beijing Rural work Committee and undertaken by Beijing Sightseeing Leisure Agricultural Trade Association, you can find 14 columns altogether. They are joyful farm households, leisure garden, festive activities, special local products, farm gourmet, scenic spots, travel know-all, a galaxy of folklore, restaurants and hotels, natural science exhibitions, leisure and body-building, collecting folk songs from different areas, association construction and tourists' forums, Here we can see they have rendered sufficient attention to the attraction and supply of rural tourism. "Joyful farm households", "leisure garden", "festive activities", "scenic spots", "a galaxy of folklore,", "natural science exhibitions" and "leisure and body-building" doubtlessly reflect the tourist interests of rural tourist consumers and "special local products", "farm gourmet" and "hotels" are listed to satisfy part supply of tourists in their tourist activities. This will surely be of great help for tourists to understand the supply of rural tourism.

However, it is not enough for tourists to generalize these. We should also have more study and exploration of the theory and practice about the development of rural tourism if we want to promote healthy development of rural tourism. It is precisely for this purpose that the author tries to make initial discriminations.

1.2 A kind of definition

Although we have had approximate consensus about the view of "rural tourism", we can also find there are still some differences about the understanding and elaboration of this concept if we follow some talks from different experts and read some of their expositions.

Is there any more agreeable explanation? If we read the definition of "Guide to Local Tourism Planning" and one of "Series of Tourism and Environment" recommended to government officials of different countries, local communities and tourist operators by World Tourism Organization, it is: "The activities of staying, studying and experiencing rural life mode in the countryside (usually traditional countryside in outlying areas) and its vicinity. That village can also be a base for tourists to explore nearby areas."

In this document, "rural tourism" is used. As is known to all, "rural tourism" is more or less the same as "agro-tourism" and "farm tourism". The former is inclined to rural amorous feelings while the latter ones can hardly be separated from farm work. Therefore, we can also translate the latter ones into "agro-tourism" and "farm tourism" [1].

1.3 Early perception in China

"Rural Tourism" developed in China in the 20th century. More care was given to it by agricultural departments in Beijing and the central government, which resulted in the establishment of China Peasant Tourism Association. In 1989, "rural tourism" was more widely focused. In April that year, the Third Convention of China Peasant Tourism Association was held in Zhenzhou, Henan Province. In accordance with the proposal of related comrades, "China Peasant Tourism Association" was renamed as "China Rural Tourism Association. (The author was listed as advisor of this association.) Owing to various reasons, "China Rural Tourism Association" was replaced by "China Domestic Tourism Association" (Afterwards it was replaced by "China Travel Service Association"). However, the official naming of "China Rural Tourism Association" reflected to a certain degree the rise of China's rural tourism in the 80s of the 20th century. (With Baidunhu Peasant Resort in Zhuhai opening for business in 1984 as main symbol) And the subsequent large-scale development nationwide reflected that rural tourism operators, rural tourism researchers and relevant departments had set their visions on this field.

1.4 Subsequent study

Despite the fact that the disappearance of "China Rural Tourism Association" resulted in the lack of approval and support in the development of rural tourism, studies among common folks in this regard slowly cropped up. Several years later, a drastic change turned up. In 1998, a nationwide symposium with "rural tourism" as its theme was jointly held in Huangshan, sponsored by China Tourism Future Research Committee, together with Tienjin Tourism Administration, Editorial Board of "Tourism Tribune", China Tourism Management Cadre Institute, Huangshan Municipal Party Committee and Huangshan Municipal Government. At the conference, delegates explored quite a number of topics concerning the development of rural tourism, combining the current situation of Zhejiang, Hunan, Tienjin and Anhui. They also made in-depth discussions on the definition of rural tourism, charms of rural tourism, resource features of rural tourism, as well as the planning characteristics, prospect and sustainable development of rural tourism and comparison of rural tourism between China and foreign countries. The conference exerted great influence among academic circle and industry at that time. However, because of the limited number of distribution of "Tourism Management" that carried these theses, the convening of this symposium failed to catch enough attention from subsequent researchers [2,3].

For instance, some authors that contributed their papers about rural tourism on "Tourism Tribune" these years did not even have chance to read that issue of above-mentioned academic journal "Tourism Management" sponsored by China Tourism Management Cadre Institute.

Although study was still going on and rural tourism was still under development, we cannot but point out that in parts of studies, there exist unclear concept and thoughts about rural tourism. That was in the 90s of the 20th century, especially before and after 1995, when more "sightseeing agriculture" was used to replace "rural tourism" in tourism industry, academic circle and tourism authorities. This concept was concentrate reflected in the book "Sightseeing Agriculture" published in 1995. The publishing of this book undoubtedly played a great pushing role in the development of China's sightseeing agriculture and rural tourism, but it also had its limitation in the initial stage of study. Therefore, in the symposium held in Huangshan in 1998, we read a thesis entitled "Is It "Rural Tourism" or "Sightseeing Agriculture?"[4] The paper made discriminations about some confusing understandings and misused expressions at that time. Just as the paper pointed out, the use of the word "sightseeing" might produce different interpretations. Before liberation, and in present-day Taiwan and Japan, the word "sightseeing" refers to "travel", which is more different from our "sightseeing-type tourism activities". And the main materials used in the book "Sightseeing Agriculture" are exactly from Japan and China's Taiwan. "Sightseeing agriculture" is only a part of agricultural production, a composition of supply in "rural tourism". The pity is a certain part of people have unclear discriminations about "sightseeing agriculture" and "rural tourism", which leads to the stagnation of the development of rural tourism and neglect of rural tourism in the past years.

What deserves our mention is that in September, 2002 the first symposium "The Development of Sightseeing Leisure Agriculture and Rural Tourism across Taiwan Straits" was held in Yanqing, Beijing. The conference was jointly sponsored by China Geography Science Society, Beijing Scientific and Technological Association, Yanqing County People's Government and an institution of higher learning in Taizhong, Taiwan.

From the title of the conference, conference summary and columns of theses collection (sightseeing agriculture and leisure agriculture, urban agriculture and modern agriculture, Taiwan rural folklore and cultural tourism, ecotourism and landscape planning, Taiwan sightseeing leisure agriculture and rural tourism, agriculture and rural economic development), we can clearly see that this conference, taking agriculture and rural areas as foothold, studied problems of developing tourism in rural areas. We can definitely say that this was really another valuable conference to promote the growth of China's rural tourism and discussed rural tourism more widely from the perspective of geography science and agricultural science, laying a new foundation for the maturity of China's rural tourism.

1.5   Key contents of rural tourism

What merits our attention is that we should explore not only the relationship between sightseeing agriculture, leisure agriculture and rural tourism from the perspective of agriculture, but also the relationship between them from the perspective of tourism industry. If we inspect from tourism activities and study from tourism discipline, the task entrusted us by "rural tourism" will not be easy. Lots of work remains to be done. The author holds the view that the key content of rural tourism should be rural amorous feelings. Rural amorous feelings seem to include the following four parts:

Local conditions and customs—particular geographical environment

Natural resources—special local scenery

Custom—local folk custom

Landscape—local landscapes that can be appreciated

The following contents are all parts that can be hard to be separated from rural tourism.

Sight—beautiful landscapes

Style and features—pleasant appearance

Graceful bearing—eye-catching demeanor and manner

Special flavor—local specialties (including local specialized food)

Elegant rhyme—folk songs, folk rhyme and folk stories

Prevailing custom—popular general mood and habit

Since the countryside has always been the place for agricultural production, rural tourism activities can be hard to be separated from farm work. Therefore, sightseeing agriculture and leisure agriculture have become indispensable supply of rural tourism. However, the contents involved in rural tourism are much wider and broader than sightseeing agriculture and leisure agriculture. Even we view from the aspect of supply; sightseeing agriculture and leisure agriculture are only a part of the supply of rural tourism (Though a particularly important part).

1.6   Definition

The author, so far as years of experience is concerned, holds the view that the definition of rural tourism is: an attraction with rural region and farm work-related amorous feelings composed of local conditions, local natural resources, local folk customs and landscapes to attract tourists to come for various tourist activities such as rest, sightseeing, experience and study. Rural tourism can be divided into traditional and modern ones. China's present-day rural tourism should no doubt include modern and traditional ones and these two parts are often inseparable. In the book "Guide to Local Tourism Planning", the definition of rural tourism usually refers to outlying traditional rural areas. We might as well make necessary revision [5].

1.7 Growth mode

In our country, the growth of contemporary rural tourism can be divided into four modes.

Tourist generating area-depended mode (or tourist generating-adjacent mode): urban residents' travel developed with the help of regional advantage close to the city (Strictly speaking, it depends on its different rural resources that are close to urban market. Its resource advantage mainly comes from natural environment whose product element has functions of sightseeing leisure, represented by "happy farm household's tour", "happy fishing household's tour" and "mountainous families' tour".

Destination-depended mode (or resource-adjacent mode): multi-tourist-generating urban and rural residents' travel developed with the help of the advantage of their own or nearby tourist attractions (Strictly speaking, it depends on the different product-mix) its resource advantages are natural environment and original scenic spots. Its product element is sightseeing and leisure. Some national minority villages in outskirts of Beijing belong to this mode. Here two cases are presented: (1) depending on special minority villages or dwelling communities (2) depending on famous natural landscapes or cultural and historical landscapes.

Non-typical mode: Although it's geographical location is in the countryside, its product mix does not belong to rural category. (The most outstanding is those urban entertaining-type resorts located in rural areas.) Its resource advantage lies in the modern innovation of natural environment. Its product element is mainly leisure. The most representative one in early years was Baitunhu Peasant Resort located in Zhuhai. Even now they are still using the following slogans for promotion: staying along the lake; playing on the water and enjoying sea food."

Compound mode: mixed formation of the above multi-modes.

From the above growth modes, we can see that it is exactly the needs of the residents (particularly urban residents) that promote the growth of rural tourism. If we inspect from the current situation, we will clearly find that in all kinds of modes regarding rural tourism, the fastest growing mode is tourism generating area-depended mode. Its product make-up is simple with more similarity, but it adapts well to the needs of the markets. (What it faces is nearly urban markets, so it breaks away from long-distance competition.) Though it still leaves room for improvement, its mode for development is beyond censure.

1.8 Adapting types for requirement

The following seven types for rural tourism are selected to satisfy the needs of tourists:

Rest and entertaining type: represented by "happy farm households tour", "happy fishing households tour" and "mountainous families tour".

Harvest and tasting type: mainly engaging in picking fruits and fishing or enjoying special cuisine and gourmet, represented by "picking fruits tour", "fishing world" and "gourmet village".

Sports and fitness type: mainly engaging in mountaineering and water sports, recuperating under rural natural environment represented by "rural sports club" and some "spa villas".

Sightseeing and sense of beauty type: mainly engaging in sightseeing travel or doing farm work, including modern rural sightseeing, scientific and technological agriculture sightseeing, ancient village dwelling sightseeing and nearby scenic spot sightseeing.

Study type: arranging travel for schools or parents, mainly for inspection, painting or sketching from nature or doing field work, represented by students' hiking and summer camps.

Compound type: It doesn't lay stress on certain type, with several types combined.

2. Rural tourism and community development

2.1 China's rural tourism has aroused global concern in recent years

At the end of 2002, Mr. Fragialli, Secretary General of World Tourism Organization and Dr. Walma, who is in charge of Asian Pacific Department, specially visited Balahe Rural Tourism Demonstration Area, Guizhou Province and gave their full affirmation to the tourism development of that area [6, 7] During World Tourism Conference in October 2003, Ministers of Tourism form various countries and nearly 300 distinguished guests form world organization listened to the speech made by director of Guizhou Tourism Administation about developing rural tourism to help the poor in Guizhou Province. By the end of

2002, 0.532 million people in the countryside had eliminated poverty through developing rural tourism. Comrades from Guizhou also put forward important measures to protect the interests of local peasants. Obviously, it is not feasible to rely on tourist enterprises alone. It is imperative to ask local people to get involved and the peasants will get rich in the course of the development of tourism. Only in this way, can the peasants be stimulated to have initiative of protecting their culture and can rural tourism achieve sustainable development[8].

2.2 Another concept Included in "Rural Tourism"

Experience from Guizhou and the concept "an effective means to help poverty elimination through sustainable development of tourism" by World Tourism Organization are worthy our serious attention. However, the concept of "anti-poverty" cannot replace tourist form, nor is "anti-poverty" a kind of tourism product. It is only one of the aims for tourism development and management.

2.3 The selection of sustainable development

Official of Department of Sustainable Development of World Tourism organization suggested seven schemes apart from expounding the advantages of tourism industry under international background and the principle of developing tourism industry to solve poverty at "The International Forum on Rural Tourism and Poverty Alleviation" held in Guizhou in October, 2004. The main contents are: 1. employing the poor in tourism enterprises to ensure their fair income; 2. providing quality and believable commodities and service to tourism enterprises by the poor people; 3. directly selling commodities and service to tourists by the poor people under orderly and easy consuming environment; 4. a system to support the poor people to establish enterprises with the arrangement of capital, technology, reputation, property right, law and marketing; 5. a plan conducive to reducing tax in poverty-stricken areas 6. Encouraging tourism enterprises and tourists to donate voluntarily 7. promoting the investment of infrastructure construction through tourism industry…… These suggestions are evidently entirely different from the practice in some places in China where rural tourism aims at increasing local revenue and profits for enterprises (The rich get richer)[9, 10].

2.4 Poverty alleviation

"Anti-poverty" measures in developing countries and China's policy of "poverty alleviation through rural tourism" are actually an act of keen insight. And for the measures of getting rid of poverty, "poverty alleviation through tourism" has the advantage of turning "blood transfusion-type" into "blood forming-type".

However, if we overstress economic target from the theory of economists, it might bring about heavy economic pressure on tourism industry. If handled improperly, it might cause great loss hard to retrieve to ecology, environment and social atmosphere of destinations. Currently, some negative impact in the course of the development of tourism has drawn forth much blame on tourist activities and tourism industry. In fact, this is not the wrong doing of tourist activities and tourism industry, most of the problems result from the inattentive consideration on the part of decision-makers or faulty operation of enterprises. The study methods and conclusions of most tourist researchers are similar to the study of anthropology, sociology, ecology, cultural science and folklore on this point.

2.5 Current relationship between rural tourism and ecology

The relationship between rural tourism and ecology (including natural ecology and humane ecology) must arouse our high attention. While making our efforts in protecting more and more scarce proto-ecology (or quasi-protoecology), we must allow different forms of ecotourism to exist.

The reason why we propose the existence of "different forms" is that we should not refer to certain proto-ecology once we mention "ecology". (Now the prevailing ecotourism internationally mostly refers to proto-ecology) The author holds the view that we should see the existence of different ecologies. They are proto-ecology; quasi-ecology; secondary ecology, regenerated ecology; newborn ecology and bionic ecology.

Therefore, ecotourism in broad sense might as well consider dividing into two categories—ecotourism and non-proto ecotourism.

1. Ecotourism: the ecological environment required is (1) proto-ecology, (2) quasi-protoecology.

2. Non-Proto ecotourism (or quasi-ecotourism): the ecological environment required is (3) secondary ecology, (4) regenerated ecology, (5) newborn ecology, (6) bionic ecology. The author suggests we add a "classified" adjective before this kind of tourism, "called......ecotourism".

In order to be consistent with the international understanding of ecotourism, we can still call it "ecotourism", "non-proto-ecotourism". In this way, we can not "run counter to" the international understanding, at the same time we can get more consensuses with other people with interest so as to avoid making general reference about "ecotourism", which might lead to serious danger of undermining proto-ecology. Travels in different types of eco-environment, as objects of different needs, are necessary. But it is very important to protect proto-ecology.

As for ecotourism with proto-ecology as carrier, the author still insists that we should formulate strict criterion for ecotourism in our country as soon as possible and achieve standardization and enforcement compulsorily, in accordance with the principle of ecotourism that is widely acknowledged and in the light of international criterion for ecotourism. Any ambiguity and appeasement (in thinking or theory) and unlimited extension of popularization will do great harm.

It should be noted here that some scholars have shed doubts about why "ecology" in "ecotourism" must be of scientific and aesthetic value that could be appreciated. Some are even dubious of the term "ecotourism". This topic is widely involved, so I am not going to dwell on it here.

Currently, opinions about the development of ecotourism in our country disagree in both academic circle and industry. Quite a few scholars think that there are still lots of difficulties in carrying out "ecotourism" that conforms to international cognition. We still have a long way to go .Considerable development has been achieved in China in terms of rural tourism, but overall it is still in the initial stage of development. However, its negative impact on rural ecology is not only of theoretical importance, but of realistic urgency. The topic "Rural Ecotourism", which combines the study of rural tourism with ecotourism by Yunnan Province, is a very pioneering topic.

Rural ecotourism, which is of very prominent significance to rural construction, not only helps preserve original ecology, but also is conducive to the further optimization and construction of rural ecology.

2.6 Opportunities and challenge brought about by the process of urbanization

With the global process of "urbanization" in human development, more and more rural areas have become "urbanized". (This has lost original rural features). This makes it more valuable to travel in rural areas and more difficult to keep original and traditional features.

Therefore, in rural travel, how to reduce the loss of rural culture and avoid "behavioral pollution" of tourists has become another important topic confronting the development of rural tourism.

3. Various forms of the development of rural tourism

In order to adapt to the needs of the market, there gradually appear respective different forms in the development of rural tourism. Followings are some most representative regional groups in present-day China.

3.1 "Happy Farm Households" in Chengdu

Although we had witnessed certain scale of rural tourism before the official birth of "China Rural Tourism Association" in 1989, large-scale of "happy farm households" have been formed in the past decade. The place of birth of "happy farm households" is in Chengdu. Its development process can be divided into three periods: spontaneous and budding period (1978-1992); popularization and development period (1992-2004); standardized period (2004-present).

The development of "happy farm households" in Chengdu is evidently based on the nearby needs of Chengdu guests. "Happy" here refers to rest, entertainment and catering.

In June 2005, happy farm households recommended by "Chengdu happy households" website occupied 28 pages (20 households per page), 548 households altogether. Estimates from various materials show that there are approximately 5000 happy farm households in Chengdu (By early 2004, registered "happy farm households" had reached 4559 [11]) at present, happy farm households (including happy fishing households and mountainous families) have at least reached 30000.

Although our researchers are trying to divide "happy farm households" into several modes and types, we cannot but see the best developed mode and most of rural tourism items that are continuing to develop are still inclined to "rest and entertaining type", because this type is most adaptable to the work rhythm, living habit and consuming level of urban tourists in our country.

Meanwhile, it indicates that rural tourism and other types are confronting more chances for development.

Though rural tourism is often called "happy farm household tour" nationwide, "happy farm households" in Chengdu are not the same. According to on-the –spot interview and inspection of the author, in the outskirts of Chengdu, quite a number of "happy farm households" are doing extra businesses of potted landscapes, nursery stocks, flowers and plants. They provide tourists with environment and catering for leisure and get-togethers (Few provide accommodations). Cards and mahjongg games are everywhere. In the outskirts of Changdu, though "happy farm households" also provide environment and catering (some provide accommodations), as well as cards and mahjongg, they also take fishing as major entertaining activities because of plenty of waters there. And in Huairou, outskirts of Beijing, apart from fishing (most fishing lovers are keen on trout and pacific herrings), providing seafood and satisfying good appetite of tourists are real purpose of both supply and demand. In comparison, in Fanyu and Zhongshan, Pearl River Delta area, despite food and beverage and fishing, what makes tourists feel wild with joy is a stretch to the horizon of paddyfields.

3.2 "Folklore Village" in Beijing

Rural tourism in Beijing can also be called "happy farm households tour" and was based on the enlightenment from "Peach Flower Festival"; but "happy farm households" are standardized with villages as symbols. Though rural tourism in Beijing started as early as in Chengdu, time for its big development is a little bit later than Chengdu. Nevertheless, its scale has now surpassed that of Chengdu.

By December 2004, folklore tourism had been carried out in Beijing's 316 administrative villages, among which 70 are municipal-level "folklore villages". Rural folklore tourist households have been developed into 13819, among which 5537 are municipal-level folklore tourist households. If we add 40 folklore tourist villages and 1582 municipal-level folklore tourist households approved in September 2006, municipal-level "folklore villages" have reached 110 and municipal-level rural folklore tourist households have reached 7119 [12, 13].

Unlike Chengdu, quite a few folklore villages in Beijing rely on historic sites or natural scenic areas, forest parks and geological parks. Moreover, if we inspect from overall perspective, folklore atmosphere of its products in Beijing is not so strong though we call rural tourism in Beijing as "folklore tourism". So it should belong to compound type in terms of its adaptive type for demand.

What merits our attention is that a study on leisure destinations in the suburbs of Beijing is now going on. Rural tourism planning in the suburbs of Beijing has been initially compiled. Counties as Miyun and Yangqing are making efforts to aim at folklore tourism. We can foresee Beijing's rural tourism is bound to witness greater development in the near future.

3.3 Ancient villages in Anhui and Jiangxi

In November 2005, two assemblies on rural tourism were quite eye-catching. One is "Wuyuan Rural Cultural Tourism Festival "held in Shangyao, Wuyuan, Jiangxi Province, the other is "Forum on China Rural Tourism" held in Yixian, Huangshan, Anhui Province. The two festivals or conferences aimed at promoting the development of rural tourism with rural settlements as center.

Yixian has tranversed twenty years' development of tourism. Its profound foundation lies in rich civilization of rural settlements with local features.

Tourism in Yixian is not only world-renowned but both Xidi and Hongcun have been listed as "world cultural heritages" by UNESCO. Located in the juncture of Anhui, Zhejiang and Jiangxi, Wuyuan lacks famous mountains and waters ,but it has won reputation of "scholarly town". Footmarks of Anhui merchants' pioneering spirit are still printed on small streets. Scores of ancient buildings built in the Ming and Qing dynasties permeate past-day flourishing atmosphere of this village. Wuyuan, which attaches great importance to preservation of culture and history, has awaken the footsteps of tourists. Not long ago, in the public appraisal of "the most beautiful place in China" by "China National Geography" and other media, "Wuyuan Ancient Village Clusters" was awarded "one of China's six most beautiful ancient towns and ancient villages".

If Xidi and Hongcun in Yixian are models of tourism development in villagers' settlements, Huangkou, Jiangwai and other unknown villages are bound to be destinations to which tourists will look forward for their profound traditional rural culture. Though in the development of rural tourism in Yixian and Wuyuan, more and more farm work and natural sceneries will be added Culture in rural settlements(both share Anhui culture) will long become important support of their tourism products. Among them, folklore culture will still be noticeable.

3.4 Rurual tourism in Taiwan

In addition, in China's Taiwan, there is also rural tourism with their own features and regional form.

The author does not have chance to inspect and lacks materials, so the description will suspend here.

4. Folklore tourism and rural tourism

4.1 Radical reform of folklore tourism

If we take Beijing as an example, we can see folklore tourism in Beijing should be divided into two branches. One branch is urban folklore tourism (represented by hutong trip), and the other is rural folklore tourism (represented by "folklore villages".

Therefore, we should not think folklore tourism only exists in rural areas. We should also see rural folklore tourism is one part of rural tourism. Moreover, since rural folklore in our country is of rich cultural connotation (It has kept quite perfectly in some areas.), it is an important part (even indispensable part) for the attraction of rural tourism. As the common people irreversibly share the same modern ways of life and tourism products of the same kind become more and more prominent, disparate folklore tourism products have become impotant knock-out products to attract tourists and strengthen competitiveness. (The attraction of Beijing's "hutong trip" in the past decade is a good example.) So when we develop rural tourism, we should pay more attention to the close relationship between folklore tourism and rural tourism. This is the exact point that the author expounded at the time when "China Rural Tourism Association" was established in 1989.

What we should face squarely is when we study folklore tourism, our interpretation of "folklore" should not be far from the cognition of folklorists both at home and abroad. What is folklore? It is an inheritance form the people. So far as the meaning of the word "Folklore" is concerned, it is folk learning. Folklore is a kind of living subculture; it is what we say "living fossil". Folklore must be obtained through the inheritance from the masses. "Pan-folklore", as a school of thought, has reason to exist. But if we replace "folklore" with "Pan-folklore", it might lead to the deviation form inheritance and it runs counter to the spirit of folklore.

4.2 The contents of folklore tourism

Traditional folklore can be roughly divided into three schools of thought:
1. humane school of thought, with folk literature and folk arts as objects of study; 2. psychological school of thought, with belief inheritance as an object of study 3. sociological school or anthropological school, with human existent folklore as an object of study.

In fact, this generally reflects three aspects of folklore.

If we merge eleven categories of folklore by late Mr. Zhong Jingwen [14] and four categories by Mr. Wu Binan [15], the contents of folklore will be reduced to three categories. 1. recreational folklore (oral literature; folk song and dance; folk music, games and folk skill......); 2. belief folklore (except for superstitions that should be cast away today, festivals and festive activities are most prominent.......); 3. productive folklore and social folklore, or economic and social folklore (productive activities, life activities and human activities——marriage, funeral, newly born, aged, sick and dead......).

In China's vast rural areas, very rich folklore can still be found and they are even kept perfectly in some areas. We have obligations to dig and protect them. Meanwhile we should strengthen our cognition of their value.

The most typical one is in shijiazhuang Village, Anqiu, Shangdong Province. In early 70s of the 20th century, Shijiazhuang was "the first village" to be opened to foreigners as "the typical representative of new socialist countryside". After the changes in late 70s, 20ᵗʰ century, with the holding of the First Weifang Kite Festival in 1984 and the opening of "Thousand-Li Folklore Tourism Route" sponsored by Shangdong Tourism Administration, this village has become world-renowned folklore tourism village in China along with Yangjiabu, home of new year paintings. Here, they have done a good job of digging folklore affairs and combined folklore with the needs of tourists pretty well.

4.3 Definition of folklore tourism

If I combine my early study of folklore with my present study of tourism, I think folklore tourism is tourism activities with folklore inheritances that closely connect urban and rural people as main tourist purpose.

In tourist activities, tourists, apart from watching, participating in and experiencing recreational folklore inheritances of local masses, belief folklore inheritances, productive folklore and people's livelihood folklore inheritances, will inevitably watch, participate in the experience the realistic life of local masses(i. e. contents referring to "pan-folklore") and non-local inheriting culture and arts.

4.4 Folklore tourism should be accorded with more attention

Our tourism operation, tourism development and planning often take "folklore tourism" as "theme" in their written and oral publicities, but in practice we do not find so many contents regarding folklore tourism. The fact is we do not understand much about the inheritance of local folklore. Besides great efforts to be made by operators, developers and planners in this field, we should welcome more folklorists to be involved in the work of tourism operation, tourism development and planning.

Here, I cannot but mention that Republic of korea succeeded in bidding "Jiangling Dragon Boat Festival" as "representative of human oral and intangible heritage" in 2005. In our country, although there is high initiative in "bidding for heritages" by local authorities, they usually keep their eyes on local cultural relics. As we notice korea, though overall Dragon Boat Festival is not a very popular festival there, people in Jiangling, Jiangyuandao, have kept this festive activity pretty well and successfully won the bidding. This should serve us a good lesson. Afterwards, some experts made comments on the media, saying that "Dragon Boat Festival in Korea is not the same as in China," "Koreans sacrifice mountain god while Chinese sacrifice water god" [16]. This shows that it is imperative for some experts in our country to study further.

With the speedy modernization of human society, folklore as life inheritance is being constantly vanished. However, a country or an area alike, rural area always lags behind urban area in the process of modernization. Therefore, folklore kept in rural areas in always more than that in urban areas. Furthermore, historic difference leads to the disparity of regional folklore features. Obviously, it is a pity that we haven't made full use of folklore affairs (abundant regional features in endless variety) in the development of rural tourism.

5. Agro—tourism and rural tourism

5.1   Official determination of agro—tourism

"Agro-tourism" was first proposed in our country in 2001. In order to conscientiously implement No. 9(2001) document and relevant leader's speech at National Conference on Tourism Development in 2001, National Tourism Administration listed industrial tourism and agro-tourism as gist for tourist work in 2001. 273 recommending materials about industrial and agro-tourism items were submitted. After analysis, investigation, selection and opinion solicitation, and discussion, the first batch of 306 exemplary lists of national industrial units were officially announced in January, 2005. 203 units (including Beijing Hancunhe) became first batch of "national agro-tourism exemplary units".

5.2   Agro—tourism promotes rural tourism more mature

Doubtlessly, China's agro-tourism started at the same time with rural tourism. The official proposal of agro-tourism was a little later. If we view from the definition of agro-tourism spot ("agro-tourism spot refers to the tourism spot with the process of agricultural production, rural style and features, laboring life of peasants as main tourist attraction") and selected tourism spots in accordance with "The check criterion of National Agro-tourism and Industrial-tourism Exemplary Units", we can see that the official proposal of "agro-tourism" lay stress on the part close to productive relation in rural tourism, so we cannot regard agro-tourism completely equivalent to rural tourism.

But on the other hand, agro-tourism has laid favorable foundation to the more extensive development of rural tourism. As main part of rural tourism, it is bound to promote China's rural tourism more mature.

In order to implement the arrangement of relevant conference of the Central Party Committee, and bring into the initiative of tourism industry better in the construction of new socialist countryside, National Tourism Administration determined that tourist publicized theme in 2006 was "China Rural Tourism

2006" and formulated striking slogan" new countryside, new tourism, new experience and new morality and custom". This will bring new opportunity for the new development of rural tourism. It will promote not only the further development of rural tourism, but also further standardization of rural tourism. It will surely play an active role in solving the problems of "agriculture, rural areas and peasants" and in the construction of new socialist countryside.

References

［1］World Tourism Organization. Sustainable development of tourism industry——Guide for local tourism planning［M］. Beijing: Tourism Education Publishing House, 1997. 55(In Chinese). 世界旅游组织:《旅游业可持续发展———地方旅游规划指》［M］, 北京:旅游教育出版社,1997, 第 55 页。

［2］Zeng Benxiang. Literature review on the Research of Tourism Eliminating Poverty in China［J］.Tourism Tribune, 2006, 21(2):89–94(In Chinese. 曾本祥:《中国旅游扶贫研究综述》［J］,《旅游学刊》2006 年第 21 卷 2 期, 第 89~94 页。

［3］Shi Qiang, Zhong linsheng, Xiang baohui. Study on the development of rural tourism in China［A］.in: Guo Huancheng, Zheng Jiangxiong. Sightseeing and leisure agriculture across Taiwan Straits and the development of rural tourism［C］. Beijing: China Mining University Pubishing House, 2004. 310–313(In Chinese). 石强、钟林生、向宝惠:《中国乡村旅游发展研究》［A］, 见郭焕成、郑健雄:《海峡两岸观光休闲农业与乡村旅游发展》［C］, 北京:中国矿业大学出版社,2004, 第 310~313 页。

［4］Liu Deqian, Fu Rong. "Rural Tourism" or "Sighting Agriculture" —Rural tourism and future［J］.Tourism Management, 1998,(4):19–21(In Chinese). 刘德谦、付蓉:《是"乡村旅游"还是"观光农业"——乡村旅游与未来》［J］,《旅游管理》1998 年第 4 期, 第 19~21 页。

［5］The same as［1］. 同［1］.

［6］World Tourism Conference: Recommending Guizhou rural tourism mode［EB/OL］. http://www.gog. com. cn, 2003–10–27(In Chinese). 世界旅游大会:《推介贵州乡村旅游模式》［EB/OL］, 金黔在线, 2003–10–27。

［7］Rural tourism: Focal point that China seeks for development and protection［EB/OL］. www. xinhuanet. cn, 2003–11–10(In Chinese). 乡村旅游:《贵州旅游扶贫的重点》［EB/OL］, 新华网,2003–11–10。

［8］Rural tourism: A balancing point that china seeks for development and protection［EB/OL］. www. xinhuanet. cn, 2004–10–19(In Chinese). 乡村旅游:《中国寻求开发与保护的平衡点》［EB/OL］, 新华网,2004–10–19。

［9］Rural tourism and poverty alleviation: Suggestions for acting plan :［EB/OL］. www. gz-travel. net, 2004–10–19(In Chinese). 乡村旅游及扶贫:《行动方案建议书》［EB/OL］, 贵州旅游在线,2004–10–19。

［10］The going up of rural tourism under the dilemma of protection and development［EB/OL］. www. sznews. com, 2004-11-01(In Chinese).《乡村旅游在保护和开发两难中加热升温》［EB/OL］，深圳新闻网，2004-11-01。

［11］"Shangri-la" in Chengdu— "Happy farm households"［N］. Sichuan Daily, 2004-09-09(04) (In Chinese).《成都的"世外桃源"——"农家乐"》［N］，2004 年 9 月 9 日第 4 期《四川日报》(服务导刊·旅游)。

［12］Communique of Standing Committee of the 12th People's Congress of Beijing(No. 21)［EB/OL］. www. bjrd. gov. cn,2005-07-22(In Chinese)《市十二届人大会常委会公报》(第 21 号)［EB/OL］，北京市人大会常委会网，2005-07-22。

［13］Beijing Rural Work Committee, Beijing Tourism Administration. The announcement regarding the lists of Beijing Folklore Tourism Villages and Beijing Folklore Reception Households［EB/OL］. www. bjxcly. com, 2005-09-24(In Chinese). 北京市农村工作委员会、北京市旅游局：《关于 2005 年北京市民俗旅游村和北京市民俗旅游接待户名单的公告》［EB/OL］，北京乡村旅游网，2005-09-24。

［14］Zhong Jingwen. An Introduction to Folklore Science［M］. Shanghai: Shanghai Literature and Art Publishing House, 1998(In Chinese). 钟敬文：《民俗学概论》［M］，上海：上海文艺出版社,1998。

［15］Wu Bingan. China Folklore Science［M］. Shenyang: Liaoning University Publishing House, 1985.11-12(In Chinese). 乌丙安：《中国民俗学》［M］，沈阳：辽宁大学出版社,1985，第 11~12 页。

［16］Korea Wins in the Bidding for Dragon-Boat Festival as UN "intangible inheritance", Chinese Experts declares that it's not bad［EB/OL］.www. nanfangdaily. com. cn, 2005-11-26(In Chinese). 联合国"非物质遗产"端午申遗之争韩国胜出，中国专家称并非坏事［EB/OL］，南方报业网，2005-11-26.

Real–life Scenery Theme: An Innovation Pattern for the Development of Ethnic Cultural Tourism

——A Case Study on the Folk Song Festival of Sister Liu on Beautiful Lijiang River

Lu Jun

(College of Social Culture & Tourism, Guangxi Normal University, Guilin 541001, China)

Abstract: "Impression·Sister Liu" is the largest realistic performance perfectly shown on the real–life sceneries of hills and water on Lijiang river. By perfectly combining the culture with the hi–tech, it is a wonderful cultural tourist masterpiece in the world. As a collection of beautiful Sceneries along Lijiang River, Guangxi Nationalities Culture and the fascinating tourist project created by top–class artists of China, it is the first ethnic scenic area on real–life scenery theme with unprecedented brand–new concept in the world and its development has conformed to the RMTP Theory for development the theme of ethnic cultural tour. This paper elaborates the cultural resource of "Impression·Sister Liu" and explains the relationship among its market, the selection of theme and tourist products. At the end, the author summarizes the referential significance of real–life scenery theme and its enlightenment to the development of China's ethnic cultural tourism.

Key words: Beautiful Lijiang·Folk Song Festival of Sister Liu; Impression·Sister Liu; real–life scenery theme; tourist development

[About author] Mr. Lu Jun, lecturer of Tourism in College of Historical Culture & Tourism at Guangxi Normal University, with research interests centred on issues related to tourism planning and ethnic cultural tourism planning.

The Scenic Area of Beautiful Lijiang·Folk Song Festival of Sister Liu is a typical Folk Song Park by using the real-life scenery in Yangshuo of Guilin, Guangxi. The attraction of the park is "Impression·Sister Liu", the biggest realistic performance perfectly shown on the real-life sceneries of hills and water in Lijiang River. By perfectly combining the culture with the hi-tech, it is a wonderful cultural tourist masterpiece in the world. As a collection of beautiful Sceneries along Lijiang River, Guangxi Ethnical Culture and the fascinating tourist project created by top-class artists of China, it is the first ethnic scenic area utilizing real-life scenery theme with unprecedented fresh concept in the world. Its extraordinary development of real-life scenery theme has shaken the world since it opened to the public in 2004 and has enjoyed a high admiration given by WTO experts, top officials from different countries and tourists home and abroad. It has been accepted as one of the first "Cultural Industrial Demonstration Bases of China" by Cultural Ministry of China.

The significance of the study of this paper is based on a new demo for development of ethnic cultural tourism on real-life sceneries, which fully conforms to the RMTP Theory. It has focused on the theme (T), which based on resource (R), taken market-oriented (M) and deepened on product (P). With a brand new viewpoint and idea it replaces the traditional way of development by a distinct theme, novelty concept and unique feature which have caused a sensation in the world.

1. Basic content of RMTP theory

The core of developing ethnic cultural tourism is carried out through the conformity and innovation around the main theme (T) that is the core and the spirit as well as the key for success. The essential target of the theme conception is to avoid or release the repeated market competition and realize a regular and detail market share. The resources (R), as the base for theme development, should serve for the theme conception no matter when it assesses the resources or transfers the product. The focus of this theme conception is the characteristic and feature of the resource, including both unexploited and exploited ones. The market (M) is the guide of theme selection, which is the compact and distillation from external time-space to the theme. The rival cooperative space and the tourists' demand analysis of the market are the flexible products of theme to lead the market and the guide and foundation of product designed to fulfill the clients' demands. The product (T) is designed according to the theme by combining and refreshing the internal and external environmental elements of natural geographic background, historical cultural tradition, social psychological accumulation and economic development level after fully analyzing the characteristics of the resource and the objective of market, completely considering external space-time combination and dividing the image and content with pertinence and uniqueness. This is the basic content of RMTP Theory (Please see Figure 1 for the details).

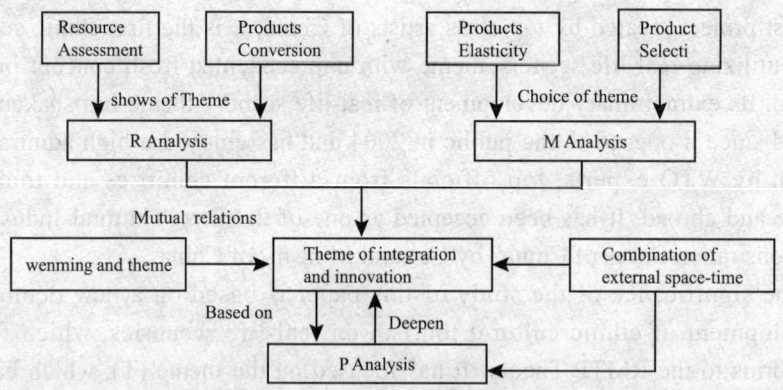

Figure 1   The Elementary Theory Frame of RMTP In Thematic Development for Ethnic Cultural Tourism

1.1   Tourist resource–foundation of the theme conception

The tourists' aesthetic tastes are various when the tourists come from different areas and ethnicities. All the tourists will analyze, appreciate and evaluate the same culture according to their own tastes during their ethnic tour. Their action will subject to and affect the psychological effects. Therefore, in the light of the tourists' different tasted demands, the ethnic cultural resources can be divided in to three theme structures: static, dynamic and abstract structures and all the tastes toward each structure along with the action of the theme and the types of theme tourist products are all different [1]. (Please see Fig. 2 for the details)

1.1.1   Static structure of resource

Located in the bottom of the structure in ethnic cultural resource, it is part of object ethnic culture. The theme tourist resources basically rely on cultural scenery resource and ethnic tourist goods, the key products of it are sightseeing travelling. In tourist aesthetics, it is in the perceptional stage, which can satisfy the tourists' aesthetic demands for natural beauties of shapes, trends, colors, structures, music and faint scents.

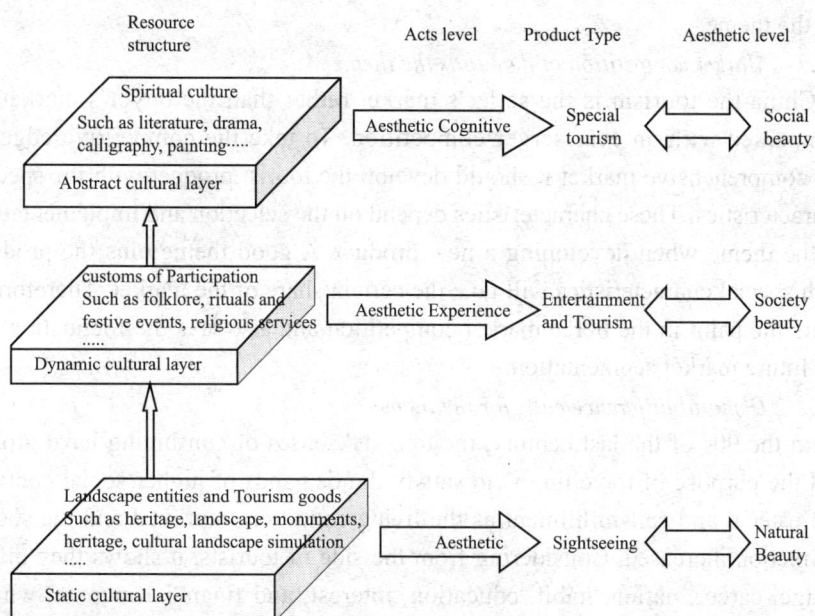

Figure 2 The Systemic Structure of Ethnic Cultural Tourism Resources

1.1.2 Dynamic structure of resource

Laying in the center (the core), the dynamic structure of resource is the folk expressions with strong trend of participation as well as dynamic and real-life culture, which belongs to folk culture of social atmosphere. It is the superiority of ethnic cultural theme tourism to keep the rival edge in severe market competition. With sharp nationality and rationality as well as strong cultural infection, it can fully satisfy the tourists' mental demand of seeking the "knowledge, freshness, difference, beauty and curiousness".

1.1.3 Abstract structure of resource

Belonging to mental culture, the abstract structure of resource is on the top of the structure. It is the core of the ethnic cultural conflict and the source of theme tourist products. It needs the tourists to experience and understand the artist charms and interacts on the ideas. It requests the tourists to posses high cultural accomplishment, certainty of aesthetic experience and attitude.

1.2 Tourism market (M): guide of theme selection

The guide of theme selection is tourism market, which lies in two aspects: the competition of the tourism market, which demands the subjects; the different market demands for the consumed demand changes of the tourists, which calls for the theme.

1.2.1 Market competition of demands the theme

In China the tourism is the seller's market rather than the buyer's market as companied with an ever server competition. To take the competitive edge in the comprehensive market it should develop the tourist product with the special characteristics. These characteristics depend on the selection and implementation of the theme when developing a new product. A good theme plus the product with special characteristics will take the certain share of the market. Therefore it bears the palm in the fierce market competition and also lays the foundation for the future market segmentation.

1.2.2 Demand difference calls for the theme

From the 90s of the last century, the tourists' senses of consuming have grown and the purpose of travelling is to satisfy the demands of higher social contact, self-esteem and self-fulfillment as the living condition improved and the social distinction increased. Considering from the side of tourists, it shows they differ in age, career, nation, habit, education, interest, and financial income, which lead to obvious differences in their psychological and physiological needs. As a result, there are various demands for tourist products which are calling for the corresponding themes.

1.3 Originality of theme (T) is the soul of tourist development

Market survey, done by multi-disciplinary experts in the aspects of ethnic culture, tourist resource and the market, is the basic job should be done in the early stage of the theme innovation. The survey on ethnic cultural resource has strongly supported to grasp the artery of the theme. To deepen the survey on tourists' resource can further refine the cultural background of the theme and provide technical prerequisite for the theme innovation. The survey and positioning of the resource is the base and premise of theme originality and the market survey and market positioning are the scientific basis and technical prerequisites for further refining the artery of the tourist resource and confirming the theme of innovation. After selecting the preliminary theme, it needs the argumentations from every aspect for the purpose of expounding and proving the feasibilities of the development scientifically. We need to collect different opinions from every aspect of the society, especially the ones from the representatives of the communities in order to ensure the maneuverability of the theme project and the equilibrium of the benefit to keep the sustained development of the theme innovation (Please see Fig. 3 for the details).

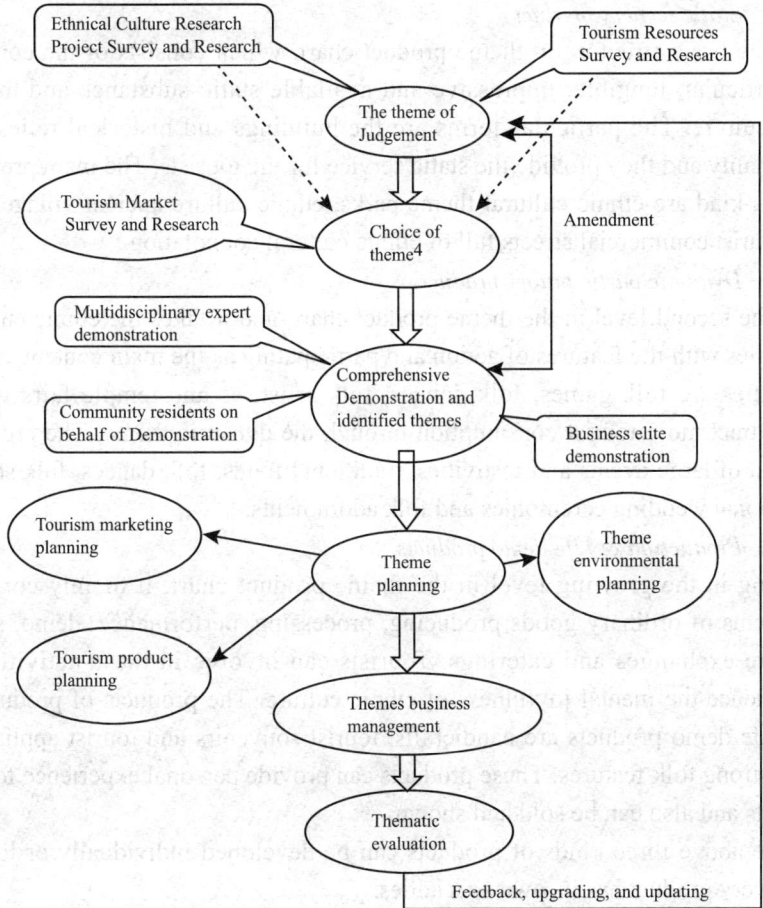

Figure 3   The general flow of the development of theme chart

1.4   Tourist product (P) is the fruit of theme innovation

As a product system, the theme product of ethnic cultural tourism is a process
to set off, create and deepen the theme. The key for designing the product is
to use the differences of the products to form distinctive and unique specialty
to satisfy the various demands for the contents of the products and form the
central competitive strength for the theme products. In the light of the layers and
differences principles of the theme, the theme product should be designed into
four-layer product chart, that is, Static scenery product, dynamic participatory
products, Production & Life demo products and Service products, which we can
refer to as SDPSL Theme Product Chart.

1.4.1 Static scenery product

It is the basic structure in theme product chart, which consists of the contents of particular, tangible, impressive and available static substance and mental folk culture. The particular forms are the buildings and historical relics of a nationality and they provide the static service for the tourists. The main products of this kind are ethnic cultural theme parks, ethnic culture ancient villages and the tourist commercial streets full of ethnic cultural connotation.

1.4.2 Dynamic participatory products

It is the second level in the theme product chart, and it takes the ethnic cultural activities with the features of action and participating as the main content. Those activities are folk games, folk dances, folk artistries and temple fairs which can attract more tourist consumption through the dynamic theme. The products consist of large events and festivities, traditional foods, folk dances, folk sports, traditional wedding ceremonies and folk adornments.

1.4.3 Production & Life demo products

Staying in the growing level in the theme product chart, it mainly contents the items of ordinary goods producing, processing, performance, demo, sales, culture exchanges and caterings. Tourists can involve in these activities to experience the mental joyfulness of ethnic culture. The products of production and life demo products are handicrafts, tourist souvenirs and tourist appliances with strong folk features. These products can provide personal experience for the tourists and also can be sold and shown.

The above three kinds of products can be developed individually or jointly and every product has its own specialties.

1.4.4 Service products

As the supplementary product, it can bring some additional value for the theme product. The main parts of the product are explication, reception and sensibilities; service is the key to ensure the quality of confirming the theme. A kind of operation culture of theme service should be formed when developing the theme.

2. (R) Analysis on cultural background of *beautiful Lijiang·folk song festival of Sister Liu*

2.1 Cultural background of beautiful Lijiang·folk song festival of Sister Liu

Sister Liu is a singing fairy according to Guangxi Zhuang Ethnic traditional legend. The film "Sister Liu", screened in Guilin, caused a sensation throughout the world in 1961 due to the picturesque sceneries of Guilin, lovely lady Sister

Liu and beautiful folk songs. At the present time, Sister Liu has become an important culture and economic brand in Guangxi and many goods, products and even the companies are named in it. *Beautiful Lijiang·Folk Song Festival of Sister Liu*, directed by the famous film director Zhang Yimou is undoubtedly the creative masterpiece by using Sister Liu as the brand.

2.2   Research on Sister Liu cultural background

Before the founding of P.R China, the research work for investigating the cultural background of Sister Liu was carried out only by some writers, artists and folklore fans. The data collected by the author shown the work scales of all dynasties were limited. After the founding of P.R. China, the culture of Sanjie Liu had once attracted more attention and "a great mass favor of Sister Liu's Culture appeared in Guangxi during 50s to late 60s of last century. Many literary and art workers rushed to collect the folk songs, the folktales and ballads to compile the materials of Sister Liu. The works about her appeared in the form of stage dramas, traditional Caidiao Opera, folk song and dance dramas, film, verses, novels, etc. Later, the film and the song and dance drama of Sister Liu had become a special art on behalf of Guangxi Zhuang Ethnic Culture to the Southeast Asian country as well as Japan." [2]

In July of 2001, the Publicity Department of Guangxi CPC organized a Sister Liu's Culture study group by the experts and professors. They went to the home town of Sister Liu in Luocheng Molao Autonomous county of Yizhou city, completed the "Research Report of Sister Liu's Culture" as the reference and held a seminar on Aug. 22 to promote the research and innovation. "During the implementation of the study, some enterprisers, planers and the top officials from local tourism bureaus were invited to participate in the project for the purpose of embodying mentality of combining the economy with the culture." [3] As a result of the long time development and cultivation, Sister Liu's Culture has become the celebrated cultural brand of Guangxi.

3. (M) Analysis on the market of *beautiful Lijiang folk song festival of Sister Liu*

3.1   Demands for cultural tourism

The comprehensive Investigation done by the National Tourism Bureau about the visitors' intend to China from the countries of the United States, Japan, UK, France and Germany shows they first prefer to learn something about the local people's life culture, which account for 100%, the second is to know the historical culture, 80% and only 40% of tourists prefer to see natural landscapes. From the statistics we know that among the tourists to the Europe, 65% go for cultural tourism. From the statistics, we can conclude that ethnic cultural tourism has been found the favor in most tourists' eyes by its unique cultural meaning and special atmosphere. Now more and more places and investors are busy developing new ethnic cultural tourist projects in recent years (Please see Fig. 4).

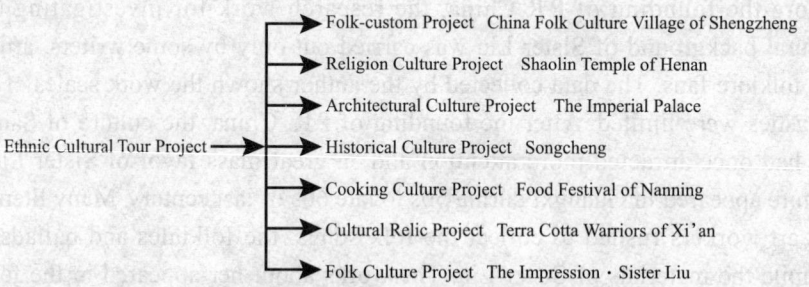

Folk-custom Project   China Folk Culture Village of Shengzheng

Religion Culture Project   Shaolin Temple of Henan

Architectural Culture Project   The Imperial Palace

Ethnic Cultural Tour Project ──→ Historical Culture Project   Songcheng

Cooking Culture Project   Food Festival of Nanning

Cultural Relic Project   Terra Cotta Warriors of Xi'an

Folk Culture Project   The Impression · Sister Liu

Figure 4   The Chart of Folk Culture Project Classification

The tourist background has laid a nice market foundation for successfully operating the project of *Beautiful Lijiang·Folk Song Festival of Sister Liu*.

3.2   Analysis of the market potential of Sister Liu's culture

The brand of Sister Liu culture has been enhanced constantly and the connotation and extension of it have been expanded to a certain extent with modernity and internationalism through many years development. The large-scale cultural events of International Festival of Folk Song in Nanning and "Impression·Sanjie Liu" of Guilin are the theme projects of Sanjie Liu's Culture developed and evaluated on the base of the "Double Three Folk Song Festival". The International Festival of Folk Song has made the brand of Sister Liu worldwide famous through the yearly performances, "especially the creative exertion in 1999's folk song festival. It mixed the pop music singing skill with the folk song performance and evoked worldwide repercussion by perfectly combining the modern performance style with traditional ones." [4] The performance of "Impression·Sister Liu" has broken down the traditional acting and stage patterns and uses the real-life sceneries of mountain and river as the stage to perform the Sister Liu's Culture incisively and vividly by using the light, color and surroundings. It has also caused a sensation throughout the world. With the modernity and internationalism, Sister Liu has gradually presented its charm and effect, thus Sister Liu's culture has entered an aero of economic pluralism, culture pluralism and open.

4. Analysis of the ethnic cultural theme (T) of
*beautiful Lijiang folk song festival of
Sister Liu* and the theme product (P)

The model of ethnic cultural theme tour will settle in a certain area or space, therefore, it should follow the geographic spatial relevance and the spatial competitive theories and rules. These spatial relevance and the competitive theoretical relations shall be applied for the following three aspects:

4.1 The spatial relevance between tourists origin and destination [5]

The development for ethnic cultural tour should be implemented in a particular spatial place, thus a spatial relevance exist between the destination of ethnic cultural tour and the tourist target market in the aspects of the spatial distance, the geographical location, the difference of culture and geographical position and the demand of the tourists to the theme.

4.2 Coordinated relation among different spatial places of the main theme

The Main Theme generally relates to many places and compared with the main theme, different places or functional subareas under the main theme will have their own significances. So we should coordinate the relationships among them.

4.3 Rivalship among the themes

At present, the whole market of the tourism is the buyer's market. As the theme can be substituted and cloned, a sharp competition for the tourists appears when developing the ethnic cultural tours in different places, in general it performs the rivals among the spatial places.

Beautiful Lijiang·Folk Song Festival of Sister Liu is the theme tourism competition with the same theme and same tourist market. In view of the effect of Sanjie Liu Brand, the scenery areas and spots by using the Sanjie Liu's Culture resources can be found everywhere in Guangxi. They are *Sanjie Liu Lascape Park of Guilin, Sanjie Liu Water Part of Yangshuo, the Park of Big Banyan in Yangshuo* and some other ones with the similar theme and tourist market. Based on this the organizer of *Beautiful Lijiang·Folk Song Festival of Sanjie Liu* had analyzed the competitors and set "Impression·Sanjie Liu" as the main theme by using bold innovation method. They had combined tow famous tourist and cultural resources of Guiling and Yangshuo as the main line of the theme, skillfully grafted and mixed the impression come from the beautiful landscape scenery of Guilin and Sanjie Liu and let the natural and cultural landscapes add radiance and beauty to each other. On the point of subtheme, they base on Guilin and the folk song tourist product of Guilin, the natural landscape and folk custom tourist product. Therefore, they have got the success.

5. Experience significance and success inspiration of theme development for *beautiful Lijiang folk song festival of Sanjie Liu*

5.1 Experience significance

5.1.1 Providing a successful model for theme development of ethnic cultural tour
Beautiful Lijiang·Folk Song Festival of Sister Liu has achieved great success in terms of its new pattern of development, bright theme and distinguished individuation. Its nice effect and experience provide a vivid model of success for the ethnic cultural development and innovation of traditional tourist spots in China. As the result, a rush of "Impression" appeared in China nowadays as *Impression West Lake, Impression Lijiang* and so on. However, it has provided a new case study demo for developing the ethnic cultural tour.

5.1.2 Changing the traditional tourist style and injecting new energy to Guilin's tourism
The *Beautiful Lijiang·Folk Song Festival of Sister Liu* is performed in the evening after enjoying the beautiful mountain and river during the daytime. This has totally changed the previous tourist pattern of doing nothing but sleeping at night after the daytime travelling. In fact the project has made the new model of travelling in Guilin: daytime for the landscape tour and night for culturescape tour.

5.1.3 Weakening the ideas of midseason and off–season
The tourist products of Guilin focus on the landscape for sightseeing, so it is with clear seasonal characteristics. But the cultural tourist product may be produced and sold in all the seasons like "Impression·Sister Liu" can be shown all the year round with its different editors in different seasons accompanied with some other cultural products. It has replaced the isolated landscape sightseeing by using its unique recreated style.

5.2 Inspiration

5.2.1 Creation—the most distinguished feature of ethnic theme tourist development
The success of the development for ethnic cultural tour depends on the efforts of creation. The development of *Beautiful Lijiang·Folk Song Festival of Sister Liu* is the result of creative process. It has broken down the traditional style of performance in the theater and built a biggest real-life scenery theater and the drum-tower group in the world, exploited Zhuang Ethnic Folk Custom Street of Dongjie as well as the real estate and commerce, which has made the creations on both contents and form. It has lasted the competitive vitality by continuously creating new product, renewing the existed ones and increasing their additional Cultural value.

5.2.2 Theme abstraction—the core of ethnic theme tourist development
The tourists' demand is varying and the folk culture is also modifiable as the society changes. The theme for ethnic cultural tour is developing dynamically with plasticity, exploitation and dynamism. It changes as the market changed, so it needs continual abstraction and distillation. The theme of "Impression·Sister Liu" of *Beautiful Lijiang·Folk Song Festival of Sister Liu* is the achievement through long-time abstraction and can stand for competition.

5.2.3 Emphasizing coordinated compromise
One of the features of Ethnic Cultural Tour development is to lay much stress on sustainable development of the theme, which includes the sustainable development of economy, environment and social benefit. The development process of *Beautiful Lijiang·Folk Song Festival of Sister Liu* has fully embodied the theory study in this paper and has enriched the content of the theory as well.

5.2.4 Stressing on community participation
The famous tourism expert Murphy said, "the amity of the place, the culture of the residents and the life style are all belong to the elements of the tourist product" [6], hence, stressing on community participation is an important content of the development of ethnic cultural tour.

Just because of the participation of the community, can the "Impression·Sister Liu" bring a real-life cultural impression to all tourists. The performance given by the local residents fully reflects the local real-lifes and the folk customs from the fishing with nets, working from sunrise, washing and wedding andso on.

References

[1] Yin Yimei. A thought on the tourism spatial competition–corporation [J] . *Journal of Guilin Institute of Tourism*, 2003, 3: 58 (in Chinese). 尹贻梅：《对旅游空间竞争与合作的思考》[J]，《桂林旅游高等专科学校学报》2003 年第 3 期，第 58 页。

[2] Nong Guanpin. Zhong Jingwen and his research of Third Sister Liu [J] . *Journal of Youjiang Teachers College For Nationalities Guangxi*, 2004, 1: 1 (in Chinese). 农冠品：《钟敬文与刘三姐研究》[J]，《广西右江民族师专学报》2004 年第 1 期，第 1 页。

[3] Huang Weilin. Guangxi tourism to the development and utilization of cultural resources on Sanjie Liu [J] . *Study of Nationalities In Guangxi*, 2002, 1: 75 (in Chinese). 黄伟林：《论广西旅游对刘三姐文化资源的开发和利用》[J]，《广西民族研究》2002 年第 1 期，第 75 页。

[4] Zhang Liquan. Building of "Liu Sanjie" cultural brand in the field of vision of cultural exchanges [J] . *Study of Nationalities In Guangxi*, 2001, 4: 59 (in Chinese). 张利群：《文化交流视野中"刘三姐"文化品牌的构建》[J]，《广西民族研究》，2001 年第 4 期，第 59 页。

[5] Li Guowen. A Research on local tourism festival planning [D] . Yunnan Normal University, Master's Degree Thesis, 2002: 21–22 (in Chinese). 李国文：《地方旅游节庆策划研究》[D]，《云南师范大学硕士研究生学位论文》，2002，第 21~22 页。

[6] Murphy, Peter E. *Tourism, A Community Approach* [M] . New York: Methuen & Co. Ltd, 1985. 215.

A Study on the Current Situation of the Development of Rural Tourism in Beijing

Wang Bing, Luo Zhenpeng, Hao Siping

(Institute of Tourism, Beijing Union University, Beijing 100101, China)

Abstract: In the recent twenty years, rural tourism is drawing the attention of the government leaders for its dramatic development and its role in alleviating the poverty situation in rural areas. Rural tourism is considered as a very important means in solving the three agriculture related issues in the national "eleventh five" plan. For this reason, the sustainable development of rural tourism is very important. Based on the survey on the operation situation of Beijing rural tourism, the service quality and visitor satisfaction on rural tourism, this paper points out problems that restrict the further development of rural tourism and suggests some issues with regard to rural tourism that need further study.

Key words: rural tourism; service quality; satisfaction; resource environment

1. Introduction

1.1 Background

With the enactment of "two-day weekend" policy in 1995 and the issue of

[About authors] Ms. Wang Bing, professor of Beijing Institute of Tourism, with research interests centred on issues related to tourism geography, tourism resource development, and rural tourism; Mr. Luo Zhenpeng, lecturer of Beijing Institute of Tourism, with research interests centred on issues related to statistics, service marketing; Mr. Hao Siping, former lecturer of Beijing Institute of Tourism, with research interests focusing on food development.

"Golden Week" policy in 1999, along with the increase of private cars, rural tourism developed dramatically all over mainland China. The most developed rural tourism areas are rural areas close to big cities and some landscapes, and the main activities are farm family experience, farm land sight seeing, fruit picking, fishing, ecological farming, farm experience, traditional cultural village, ancient village (or town, building) tour(Guo Huancheng, 2006)[1]. From July 2002 to December 2004, rules and standards related to rural tourism were issued in Beijing, Shanghai, and Chengdu respectively, and these cities became the leaders of standardized management of rural tourism in China. Academic attention to the theories and practical issues of rural tourism had been paid for the guidance of rapid development of Chinese rural tourism. In the project of "Situations and Problems of Special Food & Beverage Development in Beijing Rural Custom tourism" sponsored by Beijing Tourism Administration in 2004, we found that problems are not only limited to the development of rural Food & Beverage, but also other issues as the followings.

● The development of rural tourism was mainly in the scale of the areas in the recent 5 years; with regard to the content of rural tourism, we didn't observe significant changes, even though we call "Rural Tourism" as "Traditional Custom Tourism", we haven't seen any special traditional culture in the activities.

● Rural tourism has become more "Urbanized". Pollution on environment has become obvious with the development of rural tourism, while the natural ecological characteristics of rural areas are disappearing. Compared to developed countries, we lag behind them with a large distance.

● The unregulated private rural tourism receptions have been out of place with the enactment of varied standards and rules issued by the government, but the family based operation styles and laggard living habits changed little; as a result, there is a difference between the rules and the real practices of rural tourism operation in Beijing.

● Operators and managers of rural tourism lack the knowledge of theories on rural tourism and there is a misunderstanding of it, so they are confused in how to operate, and then they just follow what others do and how others do in rural tourism operations; as a result, we cannot see the differences between different rural tourism receptions, thus there is no special spotlight.

● Rules and quality standards are enacted, but there is no research to evaluate satisfaction and needs of tourists.

How to close the gap between rural tourism in Beijing and those in developed countries weather in terms of operation or in terms of service quality to adapt to the economy development in Beijing? What are the factors that influence the further development of Beijing rural tourism? Based on these questions, we conducted a project in June 2005, and it was sponsored by Institute of Beijing Study with the topic of *Study on the Advanced Development of Beijing Rural Tourism* and this paper is part of this project.

1.2   Objectives of the research

Through the study on current situation of Beijing rural tourism, we try to identify problems of the operations in Beijing rural tourism with the help of current theories related to rural tourism; another task is to measure service quality and customer satisfaction with rural tourism form a perspective of service management, or find out problems from the perspective of customers. We can furthermore, based on these results, pave the way for the deep development of rural tourism, and find out the resolution for farmers' problems by means of tourism; we also expect to provide evidence for scientific management and standardized monitoring in rural tourism operations; we try to identify the direction of training programs for local government leaders and farmers involved in rural tourism operations. Therefore, it is possible to provide positions for extra labor forces in local areas, to improve service quality, and to ensure the sustainable development of rural economy in a diversified way. On these purpose, our study concern the following aspects:

(1) the operation /management situation and problems in rural tourism operation in villages

(2) characteristics and behaviors of tourists in rural tourism

(3) tourists' evaluation of their experience

(4) customer satisfaction and main factors influencing customer satisfaction

(5) customer loyalty in rural tourism

2. Literature review

2.1   Rural tourism

"The term 'rural tourism' has been adopted by the European Community (EC) to refer to all tourism activity in a rural area (Roberts & Hall, 2001, p. 15) [2]." Nulty (2002. p.217) [3] indicates that the concept of rural tourism involves many elements, and the core of it is the rural community. Rural tourism relies on rural area that provides tourism places, its heritage and culture, rural activities and rural life. This concept can be outlined by figure 1. Wang Bing (1999) [4], professor of Beijing Institute of Tourism, explained rural tourism based on her study on rural tourism as "the term 'rural tourism' is relies on the resources of rural culture, rural ecological environment, farm-based activities, and traditional rural customs, involving sightseeing, researches, study, participate, entertainment, shopping, holiday as a whole tourism activity." Tourists come to rural areas to enjoy the nature, the harmony of humanistic environment and human activities; they see this enjoyment as their regression to the nature. Therefore, rural tourism has two characteristics: cultural and ecological.

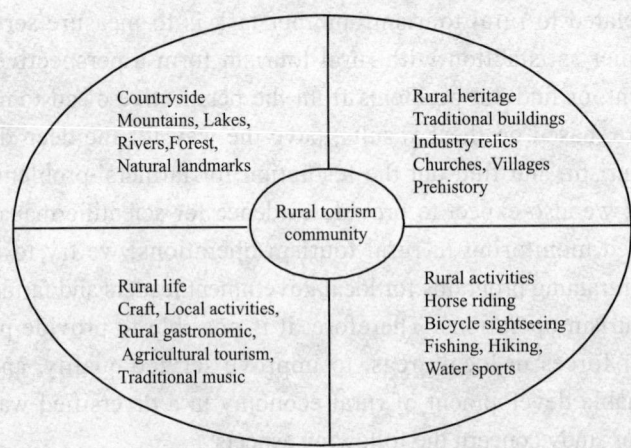

Figure 1 Concept of Rural Tourism
Source: adopted from Nulty (2002)

2.2 Objectives of rural tourism development

From the political perspective of view, rural tourism is an important means for governments all over the world to hold back the recession of traditional agricultural economy, to provide employment opportunities for farmers, to keep orders in rural societies. By providing high service quality, tourism can improve the standard of living of the local residents, provide employment opportunities, and can cause the emergence of small firms and artifact industry, so rural tourism is used by the third nations as a means of antipoverty. Chinese government proposed a slogan that "to subsidize poor farmers with tourism". How to guide the farmers to operate rural tourism formally and to ensure them to benefit from these economic activities is the key for sustainable rural tourism.

From the perspective of tourism service marketing, the purposes of developing rural tourism are: (1) to provide good services for tourists; (2) to increase revenue of the farmers through tourism services. But this two purposes should not be equally treated, only provide good services to tourists, can they attract and retain tourists, and then make profits from it.

Rural tourism without tourists is not real tourism and its development can never be sustained. Therefore, service quality play an essential role in rural tourism, and it is also the key for sustainable rural tourism.

2.3 Dimensions of tourist experience in rural tourism

According to the experience of rural tourism, services are focus on 5 factors: environment, accommodation, tourism products, special interest activities, facilities and service. These factors are the basis for measuring service quality and customer satisfaction. For measuring customers' perception of service quality, Nulty (2002, p.217) [5] provides 10 items for rural tourism as followings:

- Bed should be comfortable with reasonable price
- Bathroom should be clean
- the local food must be simple and delicious
- Enjoying local landscape, sight and the nature
- Studying local history and culture
- Experiencing holiday activities(golfing, hiking, fishing, skiing)
- Purchasing local artifacts/ souvenir and general products
- Enjoying local music, dances and drama
- No restricts on travel and the safety threaten
- Smile and friendly attitude from service employees

2.4 Measuring of service quality

According to Kandampully (2002, p.129) [6] , There are two perspectives on service quality: the operations perspective that monitor operations to ensure conformance to specifications; the other is the performance or marketing perspective that measure customer satisfaction. In this study, we conducted two surveys on the operators and on customers respectively. In measuring customer defined service quality, we measure tourists' perceptions of rural tourism services; for situation of operations, we conducted a field study in some of the villages to get the first hand information.

3. Methodology

Study 1: Survey on customer satisfaction

Measures: in measuring customer experiences, we designed 24 questions with 5-point Likert scale relating to 5 factors: *food and accommodation*, an example is "foods are special for the area": *transportation*, such as "bus system is convenient": *rural activities and shopping*, such as "we can buy local farm products": *information on rural tourism*, like "it is easy to get travel information"; and *services provided by local operators*, an example question is "villagers are friendly".

Sample and data collection: We chose 34 villages randomly within the 9 rural districts to collect data during a holiday week. 400 questionnaires were delivered by students of Beijing Institute of tourism who live in or close to the villages. 286 valid responses were received.

Study 2: Survey on operations in rural tourism in villages

Measures: We design two questionnaires to measure the current situation of Beijing rural tourism: one is for the rural tourism reception families; the other is for the village leaders responsible for rural tourism. There are 18 questions for families relating to operation situations such as "How much money was invested by your family to operate rural tourism reception?" "What is the income from rural tourism operation?" and "What is the influence by tourists expect economic issues?" questions also involved the accommodation situation, what kind of energy is used for cooking and so on; for village leaders, 25 questions involved management issues such as training, energy supplying, regulations, rural tourism activities, and facilities for operations and tourists experiences.

Sample and data collection: Among the 9 rural districts operating rural tourism, we chose 3 rural tourism villages within each district, and at least 3 to 5 farmer families were interviewed. We also interviewed leaders and committee members responsible for rural tourism in 9 typical villages. As a result, we got 63 responses from farmer families, and 34 responses from the local village leaders.

4. Findings

Findings from survey 1

4.1 Tourists behaviors in rural tourism

Table 1 shows that the main tourists are people at the age from 20 to 50 (68%), old people are very few; most of the tourists travel in one day (55%), and for the tourists travel for more than one day, most of them just stay one night (59%) at farmer's houses; with regard to the expenses, the average expense is 218 Yuan, and the variation of expenses is significant; the average travel frequency is 3 times a year, and most of them travel one time a year (50%); most travels go to rural areas with family members(49.7%) and friends (40.2%), and very few travel alone (8%) or with travel group(2%); private car is the most used means of transportation (66.1%), and secondly public bus (23.1%); the main purposes of rural tourism are tasting farmer foods, rural sight seeing, fruit picking, seeing villages, and living in farmer houses in a descending order.

Table 1　Profile of tourists

	Profile of tourist	Frequency	Percent(%)		Profile of tourist	Frequency	Percent(%)
	<20	63	22.1		alone	23	8.0
	20-29	99	34.7		friends	115	40.2
1	30-49	95	33.3	5	Family member	142	49.7
	50 or more	28	9.8		Travel group	6	2.1
	Total	285	100.0		Total	286	100.0
	No	153	55.0		farmer food	188	23.2(1)
2	Yes	125	45.0		Live in farm house	84	10.3(5)
	Total	278	100.0		Farmer land activity	46	5.7
	1 night	59	48.4		See villages	87	10.7(4)
	2 nights	33	27.0	6	Rural sight seeing	185	22.8(2)
3	3 nights	21	17.2		Fruit picking	121	14.9(3)
	4 or more	9	7.4		Health care	21	2.6
	Total	122	100.0		Fishing	68	8.4
	Private car	189	66.1		Not clear	12	1.5
	Bus	66	23.1		Total	812	100.0
4	Taxi	19	6.6		Yes	224	84.2
	Travel bus	12	4.2	7	No	42	15.8
	Total	286	100.0		Total	266	100.0

Note:

a) 1=age; 2=stay or not at night; 3 mow many nights; 4=means of travel; 5= accompanies;　6=purpose; 7= come again or not.

b) Percents are based on responses for multiple choice questions.

4.2　Tourists evaluation on their experiences

According to experiences of tourists, including foods, accommodation, travel, sight seeing, shopping, and entertainment, we design 22 related questions within the 6 factors. To avoid halo effects of respondents, we mixed the questions in the questionnaire. A final factor analysis with SPSS 12.0 shows that these 22 questions can be classified into 5 factors, and they are food and accommodation, transportation, tourism information, outdoor activities and local services. Rural tourist experience is based on these 5 factors. The results are exhibited in table 2.

Results show that the evaluation of tourists on the 5 factors is not good; each of them is less than 4.0 point, and only one of the 22 questions got a 4.01

Table 2 Descriptive statistics of tourists' experiences

	Experiences	Mean	SD	Mean	SD
1	Foods are special	3.63	.75	3.45(4)	.60
	Comfortable, clean, quiet accommodation, and there are more choices	3.57	.76		
	Bathroom is clean	3.33	.813		
	Room price is reasonable	3.72	.857		
2	Transportation is convenient	3.59	.89	3.43(5)	.77
	The bus system is good	3.40	.97		
	Taxi/bicycles, horse powered car	3.29	1.02		
3	The landscape is beautiful and natural	3.91	.81	3.52(2)	.61
	Environment(heritage, culture) is well protected without pollution and damages	3.53	.81		
	There are rich outdoor activities (fishing, water based activities, bicycling etc.)	3.47	.91		
	Night life is attractive(local dances, plays)	3.28	.96		
	Special local products are available	3.76	.90		
	Souvenir and artifacts are special	3.38	1.00		
4	Signs for directions are clear	3.62	.89	3.49(3)	.66
	Information is sufficient on this area	3.38	.85		
	Information can begotten easily	3.36	.85		
	Road situation is good	3.68	.85		
5	Hosts or employees are friendly	3.93	.77	3.88(1)	.57
	Hosts or employees appear clean and well-fitted	3.78	.76		
	Timely services without waiting	3.83	.77		
	Local residents are friendly	4.01	.71		
	Employees in tourism sights have good attitudes of services	3.82	.75		

Note: 1= food and accommodation; 2= transportation; 3=outdoor activities and shopping; 4=tourism information; 5= services.

point. In the 5 factors, service, outdoor activities, tourism information, food and accommodation, transportation are ranked in a descending order. Friend ness of local people, country sights, special rural products have high scores, but night activities and souvenir have lower scores; Tourists are not satisfied with foods and accommodation because no specialty and sanitation problems, but these problems are very crucial to rural tourism because the purposes of tourists are tasting farmer foods and living in farmer houses to experience farmer lives, so these will influence tourist satisfaction and it is reflected in the later analysis. Transportation got the lowest score both for the road to the villages and transportation in the local areas. This reflects the facilities have become a main problem in rural tourism. But this factor doesn't show significant influence on tourist satisfaction in the later regression analysis because most of them go to rural areas by private cars, but if more private cars go together, there will be a problem for satisfaction.

4.3   Tourist satisfaction and loyalty

4.3.1   Tourist satisfaction
For the question "Are you satisfied with this travel?" the average score is 3.9. This result shows that tourist satisfaction is not high or good. We can also explain it from the results of tourists' experience (table 2). The satisfied and very satisfied accounts for 75%, compared to Nulty's survey in west Ireland (tourist satisfaction with accommodation, price, foods, customer service, customer information are all over 90%), there is a big distance of tourist satisfaction between Beijing rural tourism and west countries.

4.3.2   Factors influencing satisfaction
To identify factors influencing tourist satisfaction, we conducted a regression analysis with predictors of food and accommodation, transportation, outdoor activities and shopping, tourism information, and services. Results (table 3, 4, 5) show that there is a significant linear relationship between tourist satisfaction and the predictors (sig. =0.000, table 4); furthermore, 43% of the variance of satisfaction is explained by the predictors, especially the predictors of services and food and accommodation that significantly correlated with satisfaction with significance value of 0.000 and 0.002 at the 0.05 significance level. Based on the SPSS results, we get the following regression function.

$$\text{Satisfaction} = 0.645 \times \text{services} + 0.285 \times \text{food and accommodation}$$

This function sow that tourist satisfaction with their experiences is mainly determined by services and food & accommodation. Rural tourism operator

Table 3   Model Summary

Multiple R	R Square	Adjusted R Square	S.E
.661	.437	.428	.54120

Table4   ANOVA

	Sum of Squares	df	Mean Square	F	Sig.
Regression	28.014	2	14.007	47.822	.000
Residual	36.026	123	.293		
Total	64.040	125			

Table 5   Coefficients

variable	Unstandardized		Standardized Coefficients	t value	Sig.
	Beta	Std. Error	Beta		
Constant	.502	.352		1.426	.156
Service	.645	.099	.506	6.481	.000
Food & accomm-odation	.285	.092	.242	3.093	.002

should pay more attention to these two factors to improve tourist satisfaction.

4.4 Tourist loyalty

In measuring tourist loyalty, we asked a question "Will you come back again?" 84.2% of the tourists choose 'yes', so most of the tourists have interests in rural tourism and loyalty. This also shows that the rural tourism market has good potential and this is also the basis for sustainable rural tourism development.

Findings from survey 2

4.5 Current situation and main issues in Beijing rural tourism development

4.5.1 There is no significant specialty in Beijing rural tourism

There are two kinds of rural tourism in Beijing: sightseeing (different kinds of industrialized farm lands sightseeing) and experience (visiting farmer families). We call it 'Rural Custom Tourism', and the content of it is explained by 'living in farmer houses, eating farmer foods, doing farmland work, enjoying farmer family happiness'. It is obviously that the name of 'rural custom tourism' is not consistent with the contents explained, because farmland work is only a small part of rural customs. Our study shows that people from the city are interested in fruit picking rather than farmland work and enjoy farm family happiness. Most of them like to eat farmer foods and live in farmer houses; this means what they are interested in is to enjoy the nature while tastes some thing different and find a place to relax. Why is it? Because there is a lack of needs from the market, and there is few of this kind of product and the only thing they provide is the mass education on agricultural knowledge with short life cycle in a year. The initially developed holiday tourists in rural villages are received by resorts and only some of the artists and traditional culture researchers stay in ancient cultural villages like Cuan dixia and Yao qiaoyu in Beijing.

4.5.2   The supply market of rural tourism operators is confused

(1) Farm household – a non typical market entity

The local farmers should be the operators of rural tourism according to the concept of rural tourism, but in the process of initiation and start of Beijing rural tourism, farmers are not the real initiators of rural tourism. Our survey shows that only one fourth of the rural tourism initiated by the villages themselves, and the rest was guided or promoted by local governments or the result of government promotion and farmers initiation. Generally speaking, farmers are becoming the main operators of rural tourism in terms of the scale of farmers involved in rural tourism operations, but compared to industries in the city, it is still not well organized and well developed, and this can be explained by the following reasons:

● The economy situation, life style, and standard of living of the rural farmers are still in the phase of agricultural society. Even though they are the sellers of the tourism market, they are not sensitive to market needs and their views and awareness lag behind the market changes. This is due to the long distance from the city, and the development of rural tourism is better in villages close to Beijing than those far from Beijing.

● In the process of rural tourism operation, local farmers lack the awareness of self protection as the main operators. They give the house lands to others and rent farm lands to outsiders (most of them rent the lands at least 50 years), and this ended in unbalanced competition with people from the city and other areas. As a result of it, non local operators appeared in the rural tourism market.

● Farmers lack the knowledge of tourism. Most of the operators and village leaders are not well educated, so they know little about tourism. There are training programs organized by district tourism administrations, but the contents are more about skills and regulations, so it can not cover the every aspect in rural tourism operation. In our survey, there are 33 villages were trained on sanitation knowledge, 25 villages were trained on cooking, and 26 villages received training on service behaviors. Only few village leaders received training on tourism theories, so they do not have the ideas about how to operate rural tourism.

(2) Village community – the confused main market operator

In China, rural tourism product is village based product, i.e. the product of rural tourism is defined by the domain of the village, and the name of the village is also the content of the rural tourism product – the basic unit of rural tourism product. This is promoted by different media and tourists' perceptions. Village community is one of the operators of rural tourism, and it is also the governor of family operators. But the organization of this system and the execution of management responsibility are confused. In our survey, two third of the villages have tourism association, but it is overlapped with the village committee, and the main authority is party secretary or the village leader, so tourism association in villages are not main drivers of rural tourism. We knew from the study that the founding and policies in favor of rural development from different levels of Beijing municipal government achieved very good effects, but the efficiency is low. The reasons for this are, from the macro perspective, the history, system, and policies; form the micro perspective, the essential cause if 'the lack of main entity' of rural tourism, i. e. it is not clear that 'who' is going to implement the series of policies.

(3) Competent people and rich families–strong competitors as the main operators in the market

Competent people and rich families are real operators in rural tourism receptions, and they are the following two groups: one group is people that sensitive to changes and started earlier than others, and these people are mostly village leaders; the other group is outside investors or local non farmers (invested with the money from their own companies). These rich families operate in big scale, condition of receptions is good, and normally they can accommodate 70-200 tourists each family at the same time. They can also provide entertainment activities like those in resort, but this operation is just business based, the sense of rural disappeared completely. These two groups are the main competitors for average farmer families.

(4)Animal breeding and bean curd producing households- a secondary market

We discovered that an industry chain had formed simultaneously in rural tourism operation; the tourist reception families with good places in terms of transportation have developed in big scale, and the families located in places difficult to reach but they have the appropriate environment to raise livestock and to process foods have built relationships with the former one to provide original ecological foods. We just call the later one main secondary market. This chain hasn't attracted the government's attention because this chain is not significant currently, but it reflects the influence of tourism on local industries, meanwhile, the original ecological characteristic is protected.

4.5.3   Stratified service standards and rules are urgent needs for the differences in the reception conditions of rural tourism families and the consumption levels of tourists

Beijing rural tourism has been developed in all rural districts sine 1980's. Due to the differences in location, history, and the concepts and ability of the village committee, the economic situations, reception conditions, and operation situations are different in big scales. These differences are exhibited by the following aspects: (1) the villages that close to the city, with good transportation location, relying on famous tourism attractions, have strong leaders, and as early operators of rural tourism are better than the ones lack of these advantages, even though these villages possess the original ecological environment and rich rural atmosphere; (2) the earlier operators are better than the later ones in the same village. But they use the same price in operations, so the families have good conditions are not satisfied with this pricing policy, while the families with poor condition can not retain tourists. Our survey also shows that the consumption levels of tourists are different, and they will compare the families that have good facilities, conditions and then choose the better one when they visit the village lately, so both operators and tourists call for classified standards and pricing policy, and this policy has been enacted in Shanghai and Chengdu, so it is time for Beijing to do it now.

4.5.4   Issues on the protection of energy and environment in the development of rural tourism

Resources and environment protection are old questions accompanying the development of tourism; rural tourism faces the same problems. Ancient villages are building extra houses to enlarge their capability of accommodation for tourists, and this threatens the original structure, the style and the environment of the villages and the original buildings are going to be replaced by modern buildings with new decoration and styles. Solid garbage, polluted water and the pollution of water sources are becoming serious. In new villages, traditional living habits of farmers are not consistent with the new environment and the landscape.

Energies are mainly used to prepare foods for tourists in tourism. In current situation of Beijing rural tourism, the main energies are none reproducible and environment pollution ones, reproducible energy is limited (table 5). The treatment of solid garbage is not well regulated, and most of villages use the methods of collecting, carrying it away from the village, and burying (table 6). Most of the garbage is organic materials, so some popular used treatments of organic garbage (like organically or ecologically) ways used by developed countries are not introduced into Beijing.

Table 5   Percent of energy used

Energy	Firewood	Coal	Liquid gas	Billabong gas	Electricity
Percent (%)	54.8	32.3	87.1	3.2	33.3

Table 6   Treatments of solid garbage

Treatment of solid garbage	Buried	Fired	Returned to farmland	Compost	Gas production	Total
No. of villages using the processed garbage	24	4	4	1	1	34
percent(%)	70.59	11.76	11.76	2.94	2.94	100

4.5.5   Service quality needs improvement

Based on the survey on customer satisfaction, there are following problems in rural tourism service:

(1) The development of Beijing rural tourism is in the beginning phase, so the difference in general is significant, and service quality is not good. Transportation, accommodation, and entertainment are in low quality, and foods arc not typical.

(2) Tourists satisfaction with experiences is only 75%, and average score is 3.9 with 5 point Likert scale. So tourists are not satisfied with their experiences.

(3) Tourists satisfaction is positively related to local services, foods and accommodation, and services play more important role then other issues.

(4) There are few foreigners in rural tourism market, and this is not consistent with such a big city of Beijing.

5. Discussions

The further development of rural tourism is not a new topic, but it is not uncommon that the productions of rural tourism are not significantly different among areas in Beijing, so we have the following final thoughts and suggestions on rural tourism in Beijing based on our empirical study.

5.1 Effective training of human resource

The development of rural tourism needs human resource; the development of new villages and continuous rural economy development need human resource. These two have the same purpose of development and the same purpose of human resource- local human resource. Sharpley (2001) [7] stresses that sustainable development and environment issues are the core of future rural tourism development, while the essence of sustainable development is localization. Local human resource is the key to localization. In a phase of dramatic economical development in rural areas, the primary issue is to train a generation of competent leaders to lead the development of rural tourism and future development of the new villages.

5.2 Developing diversified patterns of rural tourism

The plan of rural tourism, the exploration and exhibition of the essences of traditional culture, and the "one village, one product" design is a differentiate strategy and it needs the help of professional knowledge. While developing diversified patterns of rural tourism, it brings restriction on standards of operation, management rules, and the evaluation on levels of rural tourism development.

5.3 Standards for reception capability of rural tourism

The key to sustainable rural tourism is to keep the characteristic of rurallity, such as small size, owner of local residents, community participates, culture and sustainable environment (Brohman, 1996 cited from Zou Tongxian, 2006) [8] . Then the problems are "What is the standard for 'small size'?" "What are the related factors for the standards?" and "How developed countries restrict the number of tourists for each reception family and is it possible the use it in Beijing?"

5.4 Environment development and protection in rural areas

Based on the study on successful operational experiences in China and other countries, the key to whether rural tourism can be developed to the leisure holiday level is the environment-original or ecological like natural environment and humanistic environment. How to protect original ecology? Is it possible to build ecological like natural environment and humanistic environment? How to let local people develop modern civilization and enjoy it while building and protecting the original ecological environment? These questions are also key issues facing the sustainable rural tourism development in Beijing.

5.5 Investors from other places and their influence on rural tourism

From the theory perspective, we need to identify if industries invested by people outside the area are in the scope of rural tourism, what kind of operations could be considered to be rural tourism activities; from the perspective of empirical study, we need to answer the following questions. What are the relationship between outsiders and the local people? What are the influences of them on local people? Is it possible to coexist for the two groups and what are the bases for the coexistence? What kind of regulations should the local government put on outside investors?

References

［1］ Guo Huancheng. Developing rural tourism industry and supporting the construction of new villages in: How to develop the rural tourism with Chinese characteristics (I)［J］. *Tourism Tribune*，2006, No.3, vol. 22 (in Chinese). 郭焕成：《发展乡村旅游业，支援新农村建设》［J］,《旅游学刊》2006 年第 3 期。

［2］ Roberts L, Hall D R. *Rural Tourism and Recreation: principles to practice*［M］. New York, NY: CABI Pub. 2001.

［3］ Nulty P M. *Rural tourism in Europe: experiences, development and perspectives*［A］. UNWTO, 2002

［4］ Wang Bing. A look at the future development of Chinese rural tourism basing on a comparison of current situations of Chinese and foreign rural tourism developments［J］. *Tourism Tribune*，1999, No. 2 (in Chinese). 王兵：《从中外乡村旅游的现状对比看我国乡村旅游的未来》［J］,《旅游学刊》1999 年第 2 期。

［5］ Nulty P M. *Quality standards and training needs in rural tourism in: Rural tourism in Europe: experiences, development and perspectives*［A］. UNWTO. 2002.

［6］ Kandampully J. *Service Management:the new paradigam in hospitality*［M］. Australia: David Cunningham. 2002.

［7］ Sharpley R. *Quality standards and training needs in rural tourism in: Rural tourism and recreation: principles to practice*［M］. New York, NY: CABI Pub. 2001.

［8］ Zou Tongqian. The siege effect and treatment of rural tourism development in How to develop the rural tourism with Chinese characteristics (I)［J］. *Tourism Tribune*，2006, No. 3, vol. 22 (in Chinese). 邹统钎：《乡村旅游发展的围城效应与对策》［J］,《旅游学刊》2006 年第 3 期。

An Anthropological Analysis on the Travel of "Searching the Origin and Sacrificing the Ancestors" : the Ritualization of the Chinese Traditional Outlook on Soul[①]

Yang Lijuan[1, 2]

(1. College of Literature and Journalism, Sichuan University, Chengdu 610064, China; 2. Chengdu University of Information Technology, Chengdu 610103, China)

Abstract: From the perspective of cultural anthropology, the cultural core which is the "soul", "ghosts and gods", "root", "ancestor" in the Chinese traditional soul outlook on the tourism of searching origin and sacrificing ancestor(TSOSA) has been analyzed. After the relationship of two couples of concepts about "rite—ritualization", "tourism—ritualization" has been discriminated, That the essence of TSOSA as a social & economy phenomenon was ritualization manifestation of the Chinese traditional soul faith culture should be pointed out. Based on the tourist spiritual world exploration, the inherent motivation of TSOSA has also been proposed.

Key words: the tourism of searching the origin and sacrificing the ancestors(TSOSA); traditional soul outlook; ritualization

[Fund] This paper is supported by the key base for Humanities and Social Sciences of the Sichuan Provincial Office of Education–the research centre of protection and development on local cultural resource in 2006 (No.: 06 DFWH019).
[About author] Ms. Yang Lijuan (1978-), college of Literature and Journalism of Sichuan University 2006 doctoral candidates; Chengdu University of Information and Technology, lecturer. The research direction: anthropology of tourism and tourism economy.

1. Introduction

The worship of the ancestor is always the Chinese traditional virtue. Affected by this, the tourism of searching origin and sacrificing ancestor (TSOSA)[1] has become increasingly prosperous in recent years. Although most of the focuses are on excavating the culture of"origin and ancestor " for designing the tourism product, its understanding stays only on the specific characteristics, but without displaying the deep cultural connotation, in particular, ignores describing the tourist spirit, memory and the other scenes. So, this paper tries to explain the cultural connotation and the tourists' spiritual world of TSOSA on the perspective of the cultural anthropology, for a reasonable interpretation of the tourists' behavior based on the two compound motivations.

2. Related literature reviewing and commenting

From the perspective of anthropology, the cultural origin about tourist motivation on TSOSA is the beliefs of ancestor worship. Further more, the cultural connotations of ancestral beliefs is the traditional soul outlook. Then its essence of the development which is based on the TSOSA as the theme and the tourism as the form is the ritualization of the Chinese traditional soul outlook.

Reviewing the documents of the TSOSA, we can divided them into two aspects. On the one hand, some results focus on hot-point affection, such as Liang Liuke(1997)[1], Xu Shunchan (1998)[2], Zhao Baosheng (2003)[3], Tian Yingfang, Zheng Yaoxing (2006)[4]; the papers written by the authors mentioned above pointed out that Henan province should develop well TSOSA because of its characteristics of cultural tourism resources. Yan Lijin and Wang Yuanlin (2003)[5] have analyzed that the culture of "searching origin" is the core on the tourism culture in the overseas Chinese whose homeland is southern Fujian province, so the development of TSOSA should consider the cultural connotation and the times character of "origin". On the other hand, some authors are interested in the development of countermeasures on the TSOSA. Mao Yong (2001) analyses the favorable consumer market conditions faced

---

① The so-called "TSOSA" refers to the theme tourism whose connation is the traditional Chinese soul outlook, presenting the culture of origin and ancestors. I use Google search engine, entering Key words "TSOSA", the number of relevant information is about 64700, so that it shows it is a hot issue. The big locust tree of Hongtong in Shanxi and Yanhuang homeland is the most famous choice of destination for origin-seeking.

on Overseas Chinese in Southeast Asia with the comprehensive utilization of Confucian culture, Buddhist culture and folk culture [6]. Taking the Zuji Alley as an example, WU Liangsheng (2003) has analyzed the value exploitation can't remain on the surface layer when we take the TSOSA as the cultural tourism. The tourist destination should carry out a full excavation, and identify if the culture is compatible with the TSOSA, eventuallyt all tourism resources should be integrated into one comprehensive product, enhancing tourism levels [7]; YI He (2005) thinks that the Hakka culture features are embodied in Meizhou of Guangdong, Tingzhou of Fujian, Ganzhou of Jiangxi. The "three places" has increasingly become a popular place for the TSOSA which attracttourists at home and abroad, so they should be deeply exploited from the perspective of Hakka culture [8]; Some scholars have also analysed the tourism development on the culture of the Yellow Emperor which is affected by the prosperity of the TSOSA from overseas Chinese in the coordinated aspect, so "abandoning disputes and exploiting in common" is the most sensible choice for the integration of Cultural resources on the Yellow Emperor in every place (Shi Zhenqi, 2005) [9]. Wang Qingyun (2002) [10], Zhu Yueshun (2002) [11], Chen YanHong (2002) [12] put forward a method to "search origin" from local records, files and records of clan from the technical point of view of literature, therefore promoting the TSOSA.

It is not difficult to find, for the TSOSA, the exploitation research is the focus but the basic theoretical research is in severe shortage. In fact, the rational concern on the cultural connotation and tourist spiritual world of the TSOSA is valuable to the deep understanding of the subject and object in this kind of tourism, also to the effective guidance in the form of designing and marketing, both of them having basic theory value.

2.1 The core culture of the TSOSA: the Chinese traditional soul outlook

The beginning, existing and ending of the life are affected by "Hun-Po"[1], even the entire life course is closely related to the "Hun-Po", also "three Hun and seven 'Po'" concept is popular in folklore and the ideology of the Chinese people has long been deeply imprinted by the national culture. Modern science can not prove that the existence of "Hun-Po" and it's the dross of the superstition in the contemporary people's concept. But in my view, ignoring

---

[1] Hun-Po, which is the Chinese traditional soul outlook, 《The Ancient Explanation of Words》 explains: "Po is the negative soul; Hun is the positive soul. " Just like the "yin" and "yang" in the Taoism system. In the folk thought, there is a popular saying "the people's soul is made up from three Hun and seven Po. "

the Superstition elements of the feudal ideology "Hun-Po" is the symbol of the individual's life, the rational thinking about the other side of the world, or even the representative of the spirit of the entire nation groups, such as the national soul. The TSOSA refers to the tourism product whose theme is about searching origin and sacrificing ancestor in this paper.

The implication of the "root and ancestor" culture core in the TSOSA is the Chinese traditional soul outlook which is the integration of the Chinese traditional concept of life and ethics.

The connotation of the Chinese traditional soul outlook reflects the ancients' understanding of life processes. "Hun-Po" become a part of life after getting into the human body and plays their roles[1]; "Hun-Po" will leave the human body and change into the spirits[2](ghosts and gods) at the end of life. Spirits are not a part of the real world, so people have imagination and fiction on the spirit characteristics, even think that the spirits have a strong force, holding a fear and respect feeling toward them. People will be blessed if offering a sacrifice to the spirits with a rite. When the spirits are not ordinary, but have blood ties with the living, such as relatives and ancestors, the spirits worship is transplanted into the ancestor worship. It is said that the "Hun-Po" of the ancestors will be immortal forever and have tremendous power for offspring to protect and bless their destiny. Therefore, the ancestor worship become more popular which is based on the "Hun-Po" and the spirits worship. The focus of the ancestor worship is looking for the origin and the rite of ancestors, and this action expresses the reverence of the descendants for their ancestors.

The Chinese traditional soul outlook is the inter motivation of the TSOSA, the origin, ancestor and the "Hun-Po" are closely related to each other through the modern form of travel. Primary meaning of the origin[3] refers to the root of plant which absorb nutrients through the water or soil, and later extended to the start of things, the foundation and basis. Ancestors [4]refers "Shi", "Shi" is

---

[1] Kong Yingda interprets 《Zuo Zhuan》 : Po is the shape of the souls, Hun is the spirit of the soul. "Po is in charge of eyes, ears, heart, hands, feet and movement; the soul is in charge of spirits and consciousness. "

[2] "Li ji" records: Zaiwo(person' name) said: "I heard of the saying of ghost and god but I wonder if it is so-called. " Confucius said: "Spirit is the gathering of god. Po is the gathering of ghost. " Here God is made up from Hun and ghost is made up from Po. "Li ji" said: "Hun will reach the sky and Po will reach the earth. " This shows that after death, Hun will reach the sky and change to god; Po will reach the earth and change to ghosts.

[3] "The Ancient Explanation of Words": Root, the bottom of the tree. Its symbolic significance is the beginning and origin of the thing.

[4] "The Ancient Explanation of Words": ancestor, the temple to worship the predecessors.

related to the rite and clan temple; its primary meaning: ancestral temple[①], but also expand its meaning to ancestors, which means everyone above since his grandfather generation[②]. "Hun-Po" means "yin" and "yang" which is translated in the former. It is concluded in the TSOSA that: "Origin" is the most basic, which is signified about an abstract feeling of returning home from people's inner hearts. "Ancestor" is the specific image of the "origin", and is a signifier about the abstract feeling of "origin", which has blood ties with the people who search the origin and scarifice the ancestor, the signifier also is a specific carrier about the feeling of returning home. "Hun-Po" is a feeling signal of the changing about returning the "origin" to the "ancestors", because people believe that the "Hun-Po" of the ancestors is rooted in their homeland, so that when finding their "origin" they start sacrificing the "ancestor". It embodies their emotion of returning home and also their own identity.

With the spread of about provincialism of the Chinese people, the extension of the origin, the ancestor and the soul also will be expanded: the origin from the family to the family clan to the nation; the ancestor from the family, to the family clan, to the nation; the soul from the family, to the family clan, to the nation. It is known that origin, ancestor and soul have a close link, the decoder of the soul outlook embraces signified "origin" and signifier "ancestor", so the Chinese traditional soul outlook plays a built-in connections and guidance role.

2.2 The Proposition of the "TSOSA" and Ritualization: the Analysis Based on the Two Pairs of Relations

The traditional rite of searching origin and sacrificing ancestor is a return of spirit which go through in awe the domain of life and death of ancestor's "Hun-Po", and the "TSOSA" is a contemporary ritualization of traditional soul outlook, which both contains the traditional elements of "searching origin and sacrificing ancestors" and modern elements of "travel".

From the searching origin and sacrificing ancestors to the "TSOSA", we call it the ritualization process of traditional soul outlook. Here, two pairs of relations must be clarified: one is rite and ritualization, the other is travel and ritualization.

The rite is a kind of fixed programmable form of behavior, and has been seen as a particular act and a social practice [13], so the rite of sacrificing ancestors is a kind of fixed programmable form of behavior. The discourse symbols of "soul awe" expresses through the fixed behavior and composes the corresponding

---

①　"Zhou Li": "the ancestor temple is on the left and the temple of the land of god is on the right"
②　"the book of Poetry": you can't forget the virtue of your father.

social values, such as "filial piety". This is the rite connotation of searching origin and sacrificing ancestors, and the programmable behavior is the explicit of rite. Ritualization is a variety of rite, which changes with the passage of time and people's understanding. Compared with the rite of searching origin and sacrificing ancestor, the TSOSA reserves the rite connotation: the explicit of social values, which leads the single programmable behavior to various form. The variety which reserves the connotation and changes the form is the contemporary expression of the rite of the searching origin and sacrificing ancestors.

In the tourism activities, "trip" refers to displacement from one place to another; and "travel" refers to sightseeing in the destination, so "travel" is the core and "trip" is only the means. When traveling, people can escape from the complicated daily world, seeking some kind release of psychological tension "When confused with the daily life, people can get a profound adjustment, treatment, rehabilitation and enhancing on psychology"[14], realizing the change from the daily world to spiritual world. Therefore, people get a certain tension when far away from the homeland, missing the ancestor, getting rid of chores and returning of peace. All kinds of activities in the traditional rite of searching origin and sacrificing ancestors, such as option of auspicious day, bath before the rite, sacrifice, burning incense to call the spirits, priest leading and wearing the worship clothes etc. can't be strictly "authentically reappearance". When inducting the tourism mode, the activities of"searching origin" and "sacrificing ancestors" become the "perfect" behavior which composes of the real place and real scene. This "ritualization" innovation of "travel", in Graburn's view ①(Graburn, 1989), the intervention of "travel" make ritualization pilgrimage possible which help people from daily original status to spiritual holy status and then to daily renewal status.

2.3 The "TSOSA": the deconstruction of the ritualization presentation of the traditional soul outlook

From the single rite of searching origin and sacrificing ancestors based on the traditional soul outlook to the modern ritualization presentation of searching origin and sacrificing ancestors, two main lines are shaped in the course, one belongs to "religious system" and the other belongs to "ethical system", which

---

① In the Graburn's(1989)view, the process which includes the unusual sacred peak point before the tourists landing on the normal, secular, the general state of the existence is similar with the transition rite of human life including the pilgrimage rite. (Source: Denison ? Nash. Zong Xiaolian in translation. Anthropology of Tourism [M] . Kunming: Yunnan University Press.2004 in April edition 1: 38.)

are held together by time and developed in parallel. At last, the variety from "rite(searching origin and sacrificing ancestors)" to "ritualization(searching origin and sacrificing ancestors and travel)" comes true through the tourism. The traditional soul outlook of contemporary tourists is presented through the "TSOSA"

The traditional soul outlook is the link of the origin and the ancestor, and the feeling of Chinese people for the origin and the ancestor, is sacred presenting awing religion value. So searching the origin and sacrificing the ancestors is extremely devout rite to show people's awe, the fixed programmable behavior are "searching" and "sacrificing". The searching is searching the origin of own, one's family, one's clan and the nation. Such as the surname origin of genealogy which is the searching of the origin of the family and clan; various dynasties in history searched of the homeland of Yan and Huang Emperor several times, which is the searching of origin of nation. "Sacrificing" is sacrificing the ancestors of family, clan, and nation after finding the origin, such as sacrificing the ancestors of the surname, Yan and Huang Emperor. The sacrificing is often held before the new-year or when the family, clan or nation experiences important events, "a major event in the country is war and sacrificing". The people who emcee the rite is the fulltime Priest of the clan or the court with strict requirements which can not be changed. The main purpose of "searching" is radically reform, finding the origin's correct position, preparing for the sacrificing. The main purpose of "sacrificing" is to sacrificing and praying for the soul of the ancestors through the rite of sacrificing, then the blessing of great force can help them to realize their desire.

At this time, "searching" is the symbol of "awing" and "sacrificing" is the symbol of "praying". "Awing" and "praying" are the symbol of the social value which belongs to religious awe System.

With the improvement of the production tools and technology, a highly industrialized and civilized modern society has already arrived. People's religious consciousness is gradually reduced, the devout religious beliefs of the "soul", "ghost and gods" and ancestors are also fading.

The scared feeling of the origin and ancestors changes to the close feeling in the traditional soul outlook. On this basis, the close feeling embodies filial piety with the religious system turning into the ethical system. The "TSOSA" is the ritualization of the traditional soul outlook combining tourism. The behavior of "searching" and "sacrificing" has been arranged in the "travel". At this time, the "searching" is also the searching of the origin of oneself, family, clan and nation, such as Lien Chan and James Soong's coming to Xi' an looking for the ancestor graves in 2005,; people's visit to the big locust tree of Hongdong in Shanxi, is looking for the origin of the clan; overseas Chinese' visit to the Shaanxi is looking for the Huang Emperor grave, finding the origin of the nation. "Sacrificing" is sacrificing the ancestors of family, clan and nation after finding the origin. Such as the sacrificing of Lien Chan for his mother; people for big locust tree of Hongdong in Shanxi; overseas Chinese for Huang emperor. The "sacrificing" is held on the birthday of the sacrificing objects or according to people's time for travel. The priest who emcees the rite is played by ordinary people or local administrative officials. At this time, the main purpose of "searching" is finding the blood line destination, emotional dependence; "sacrificing" is a bailment of missing for the ancestors. "Searching" is the symbol of the "visit" and "sacrificing" is the symbol of "missing" "visit" and "missing" is the symbol of the social value of "filial piety" as the ethical system.

The traditional soul outlook mold the sanctity of the origin and ancestor, the awful religion feeling for the origin and ancestor is embodied in the rite of "searching origin and sacrificing ancestors", but people's religious consciousness has gradually weakened with the passage of time and progress of society, so the feeling for the origin and ancestors changes from the awful religion psychology to the close filial piety. Although the inherent social value changes, people still continue the rite of "searching origin and sacrificing ancestors" by means of travel, and the rite is only a ritualization behavior. So, the cultural connation of the "TSOSA" is the spiritual course of the ritualization on the traditional soul outlook, people finish the spiritual pass of "awing and praying" for ancestors through "visit and missing" process.

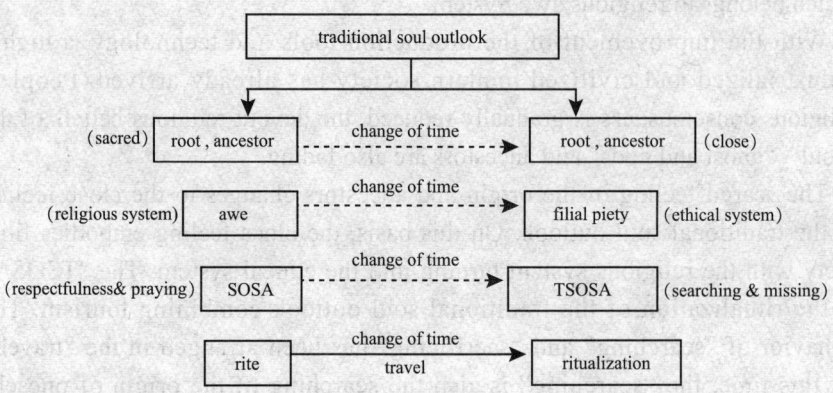

3. Conclusion

The TSOSA attracts tourists with its unique culture about origin and ancestor, in essence, it is the ritualization show of the Chinese traditional soul outlook. Analyzing the tourists from the cultural psychological, the traditional soul outlook is deeply rooted in the memory of the spirit of the people. Even people are aboard, they miss the "origin" and "ancestor" so much from day to night, thus become the potential or practical tourists who realize the searching origin and sacrificing ancestor as the modern pilgrims through the travel. Modern people complete the ritualization return of the "origin" and "ancestor"

Acknowledgement: Thanks to my tutor Pro. Xu Xinjian's for his careful guidance, as well as the valuable advices of Dr. Lai Bin.

References

[1] Liang Liuke. The research of development on special topic tourism resources in Henan [J]. *Henan University Journal* (Natural), 1997, (1) :71–76. (In Chinese). 梁留科:《开发河南专题旅游资源初探》[J],《河南大学学报》(自然版),1997 年第 1 期，第 71~76 页。

[2] Xu Shunzhan. Economic revitalization of Henan through on the advantage of cultural relic—origin tour is the focus of the tourism industry in Henan [J]. *Zhongyuan Heritage*, 1998, (4) :111–114. (In Chinese). 许顺湛:《发挥文物优势振兴河南经济——寻根游是河南旅游业的重点》[J],《中原文物》1998 年第 4 期，第 111~114 页。

[3] Zhao BaoSheng. The characteristics of Henan tourism resources and its impact of tourism [J]. *The Journal Cadres of Economic Management Institute in Zhengzhou*, 2004, (2): 24–26. (In Chinese). 赵保生:《河南旅游资源特色及其对旅游活动的影响》[J],《郑州经济管理干部学院学报》2004 年第 2 期，第 24~26 页。

[4] Tian Yingfang, Zheng yao–xing. The Development of origin tourism in Henan [J]. *The Journal of Henan University of Science and Technology*, 2006, (4) 82–85. (In Chinese). 田迎芳、郑耀星:《浅析河南寻根旅游开发》[J],《河南科技大学学报》2006 年第 4 期，第 82~85 页。

[5] Yan LiJin, Wang Yuanlin. The origin tourism of overseas Chinese homeland of southern Fujian [J]. *Human Geography*, 2003, (6) :48–51. (In Chinese). 颜丽金、王元林:《闽南侨乡寻根旅游之探讨》[J],《人文地理》2003 年第 6 期，第 48~51 页。

[6] Mao Yong. Thought of our country's attempt to develop the origin tourism of overseas Chinese in south–east Asia [J]. *Journal of Guilin institute of tourism*, 2001, (3) :72–75. (In Chinese). 毛勇:《对我国发展东南亚华人华侨寻根旅游的思考》[J],《桂林旅游高等专科学校学报》2001 年第 3 期，第 72~75 页。

[7] Wu Liangsheng. Deep development of origin tourism—a case study of Zhuji lane [J].
Journal of Shaoguan University, 2003, (5) :70–72. (In Chinese). 吴良生:《刍议寻根
旅游的深度开发—以珠玑古巷为例》[J],《韶关学院学报》2003 年第 5 期，第
70~72 页。

[8] Yi He. Culture tourism of Hakka three places, [J]. *Cross–strait relations*, 2005, (5):64–66.
(In Chinese). 一鹤:《客家三州文化游》[J],《两岸关系》2005 年第 5 期，第
64~66 页。

[9] Shi Zhenqi. The development and integration of the Huang Emperor cultural tourism
resources [J]. *Economic and Social Development*, 2005, (10) :72–76. (In Chinese). 石
振奇:《论黄帝文化旅游资源的开发与整合》[J],《经济与社会发展》2005 年第
10 期，第 72~76 页。

[10] Wang Qingyun. The Chinese ancient scenic literature and its modern value [J].
Journal of Yantai University (philosophy and social sciences), 2002, (4) :420–422. (In
Chinese). 王庆云:《中国古方志景观文学及其当代增值》[J],《烟台大学学报》
(哲学社会科学版)2002 年第 4 期，第 420–422 页。

[11] ZHu Yueshun. development of tourism resources on the searching origin through the
dossier investigation [J]. *Lantai World*, 2002, (1): 28. (In Chinese). 朱月顺:《查档
寻根开发旅游资源》[J],《兰台世界》2002 年第 1 期，第 28 页。

[12] Chen Yanhong. The research of pedigree dossier in the role of origin tourism development
[J]. *Hunan files*, 2002, (6) :13–14. (In Chinese). 陈艳红:《浅析家谱档案在寻根旅
游开发中的作用》[J],《湖南档案》2002 年第 6 期，第 13~14 页。

[13] Peng Zhaorong. *Anthropology of Tourism* [M]. Beijing: Nationalities Publishing House,
2004. (In Chinese). 彭兆荣:《旅游人类学》[M]，北京:民族出版社，2004。

[14] Zheng Qingyun. *The essence of the spiritual and cultural on tourism* [A]. Zhang
Xiaoping. *Anthropological Perspective of the National Tourism* [C]. Kunming: Yunnan
University Press, 2005. (In Chinese). 郑晴云:《论旅游的精神文化本质》[A]，张
晓萍:《民族旅游的人类学透视》[C]，昆明:云南大学出版社，2005。

Afterword

This *Chinese Tourism Research Annual 2008* in your hands is the third book of Chinese Tourism Research Annual.

It is our long wish to publish an English edition to foreign readers. thanks to the following people, who have helped us to fulfill the dream: authors of the 20 articles in this edition, scholars who are the selection and leaders of the Institute of Tourism, Beijing Union University, who are the sponsors of *Tourism Tribune* *also*.

The translations for the 20 articles in this edition was finished by the original authors and revised by experienced English teachers from the Department Foreign Language and Tourism Culture Institute of Tourism, Beijing Union University. However, as the time for translation is limited and urgent; and moreover, in this edition, maybe you will discover some non-native usage of words, we hope readers don't hesitate to tell and help us perfecting the edition.

Editorial Department of *Tourism Tribune*

August, 2008

审 稿 专 家

English Revisers
(in alphabetical order)

Chen Jie	Huang Yuhong	Liu Aifu
Pan Suling	Yuan Lihui	Zhao Li

校 译 专 家

（按英文字母顺序排列）

陈 洁	黄育红	刘爱服
潘素玲	袁立辉	赵 丽

Editors

Executive Editor-in-Chief   Song Zhiwei
Executive Deputy Editor-in-Chief   Lian Yuejuan
Executive Editors(in alphabetical order)

Lian Yuejuan   Song Zhiwei   Song Ziqian
Wu Qiaohong   Zhang Xianyu

本书编辑成员

执 行 主 编　宋志伟

执行副主编　廉月娟

执 行 编 辑 （按英文字母顺序排列）

廉月娟　宋志伟　宋子千　吴巧红　张宪玉

Institute of Tourism, Beijing Union University

Beijing Union University is the first college focused on tourism education in China, which with complete four-years and three-years programs.

Since its establishment in 1978, it has set up extensive academic exchanges with its foreign counterparts based on advanced study programs for teachers and students. For example, exchange programs with tourism colleges in France, USA, Britain, Netherlands, Norway and South Korea, which have been proved to be a two-win approach for both the teachers and students. Moreover, there are many national and international level tourism symposiums have been hosted by BIT or co-hosted by BIT and its foreign counterparts, which with the intentions to enforce the developments in tourism research. BIT also founded an institute for tourism research and the publication *Tourism Tribune*, which is an authoritative periodical and has been evaluated for four times as a "Centered Chinese Periodical" in national level. The publication for the English edition of Tourism Tribune provides a chance to strengthen its relationship with tourism research institute in all over the world.

北京联合大学旅游学院

北京联合大学旅游学院，是中国成立最早的培养本、专科旅游管理人才的高等院校。

自1978年创立以来，学院面向和服务于旅游业，首创了中国旅游人才培养的专业模式和符合学科要求的课程体系。

学院不仅是中国众多旅游专业教材的产生地，也是中国旅游研究的重要基地。学院创办的《旅游学刊》，一直是中国最有影响的权威刊物，并在中国最有学术地位的全国中文核心期刊的鉴定中，连续4次被评为全国中文核心期刊。同时，学院还依托《旅游学刊》经常举办各种主题的全国性旅游学术会议，从而对中国旅游学科的进步和中国旅游业的发展产生了十分广泛的影响。

旅游学院与国外同类院校开展了广泛的交流，除了友好的互访和联合举办国际学术会议外，学院的师资进修和学生的教育已逐步纳入国际合作的轨道，并先后与法国、美国、英国、荷兰、挪威和韩国等国的旅游院校建立了多种成熟的学生交流模式。

北京联合大学旅游学院，愿意继续为推动中国与世界的旅游学术交流再尽一份自己的力量。

About Tourism Tribune

The Tourism Tribune, an authoritative periodical in tourism research, has published 150 issues since its first launch in 1986.

As the demands in travel industry is increasing, *Tourism Tribune* has developed from a quarterly, then a bimonthly, and to a monthly issuance now. the manuscripts of this Chinese periodical (with English titles and abstracts) are mainly from China's mainland, Hong Kong, Macao as well as other overseas countries while its readers include tourism researchers, travel industry managers, government officials, teachers and students of tourism colleges.

The publication for *Tourism Tribune* is chiefly organized by Institute of Tourism, Beijing Union University, The chief editor was held by the president or vice president of Institute of Tourism, Beijing Union University by turns. The present chief editor is Mr. Zhao Peng, currently hold the post of president of Institute of Tourism, Beijing Union University

Tourism Tribune is managed by tourism experts with a senior academic title of doctorial or master degrees. The present members of the editorial department are associate professor Song Zhiwei (deputy editor–in–chief), Miss Wu Qiaohong (deputy editor–in–chief), Lian Yuejuan (M. A.), Song Ziqian (Ph. D.), Zhang Xianyu (Ph. D.), Miss Wang Yujie, Miss Zhang Fenghong.

旅 游 学 刊

《旅游学刊》是中国最受学界、业界欢迎并最有影响力的旅游专业期刊和旅游学术期刊。

自1986年创刊以来，《旅游学刊》已出刊150期。由于稿源的充足和读者的欢迎，所以他从创刊初期的季刊改为了后来的双月刊。为了适应旅游学科的进步和旅游产业的大发展，2006年又改为月刊。

《旅游学刊》以中文作为刊物的第一语言（同时刊出对应的英文标题和内容摘要）。稿源来自中国全国各地（包括中国的港澳台地区），同时也有不少海外作者向刊物寄来中文稿件。它的读者对象是旅游研究工作者、旅游实践工作者、旅游管理机构的官员和旅游高等院校师生。

《旅游学刊》由北京联合大学旅游学院主办。自创刊以来，学院的院长或副院长轮流兼任该刊的主编。现任主编是现任院长赵鹏先生。

《旅游学刊》的编辑工作是由旅游学科和相关学科的专家承担的，其中除具有中国严格制度下的高级职称任职资格外，年青同志大都具有博士或硕士学位。目前编辑部的组成是：副主编：宋志伟副编审（副教授），吴巧红女士；编辑部成员：廉月娟硕士、宋子千博士、王玉洁女士、张凤红女士、张宪玉博士。

Tourism Experts of Selected Articles for this Edition

(in alphabetical order)

Feng Dongming   associate professor, vice president of Institute of Tourism, Beijing Union University

Lian Yuejuan   editor of *Tourism Tribune*, Institute of Tourism , Beijing Union University

Liu Deqian   professor of Institute of Tourism, Beijing Union University, hornoring editor–in–chief of *Tourism tribune*

Luo Xuhua   professor, the dean of the Department of Hotel and Catering Management, Institute of Tourism, Beijing Union University

Ning Zequn   professor, the dean of the Department of Leisure Studies and Tourism Management, Institute of Tourism, Beijing Union University

Song Zhiwei   associate professor, deputy editor–in–chief of *Tourism tribune*, Institute of Tourism, Beijing Union University

Song Ziqian   Ph. D. and editor of *Tourism Tribune*, Institute of Tourism, Beijing Union University

Wang Bing   professor of Institute of Tourism, Beijing Union University

Wang Meiping associate professor, vice president of Institute of Tourism, Beijing Union University

Wu Qiaohong   deputy editor–in–chief of Tourism Tribune, Institute of Tourism, Beijing Union University

Zhao Li   professor, the dean of the Department of Foreign Language and Tourism Culture, Institute of Tourism, Beijing Union University

Zhao Peng   professor, president of Institute of Tourism, Beijing Union University

Zhu Xiyan   professor of Institute of Tourism, Beijing Union University

图书在版编目（CIP）数据

中国旅游研究年刊. 2008 = Chinese Tourism Research Annual.
2008：英文/《旅游学刊》编辑部编. —北京：社会科学文
献出版社，2008.12
ISBN 978 - 7 - 5097 - 0536 - 0

Ⅰ. 中… Ⅱ. 旅… Ⅲ. 旅游 - 研究 - 中国 - 2008 - 年
刊 - 英文 Ⅳ. F592 - 54

中国版本图书馆 CIP 数据核字（2008）第 194584 号

Cataloguing in Publication Data

Chinese Tourism Research Annual 2008

Tourism Tribune (2006 - 2007) English Edition

Editorial Department of *Tourism Tribune*

December 2008

ISBN 978 - 7 - 5097 - 0536 - 0/F · 0188

Published by Social Sciences Academic Press (China)

No. 10, Xianxiao Bystreet, Dongcheng District

Beijing, China

Postcode：100005

© 2008 by Social Sciences Academic Press (China)

Editor：Zhu Debin   Yin Pumin

Cover designer：Cai Changhai

Printed in Beijing, China